# MACHINE LEARNING FOR PHYSICS AND ASTRONOMY

# MACHINE LEARNING FOR PHYSICS AND ASTRONOMY

VIVIANA ACQUAVIVA

PRINCETON UNIVERSITY PRESS
PRINCETON AND OXFORD

Published by Princeton University Press
41 William Street, Princeton, New Jersey 08540
99 Banbury Road, Oxford OX2 6JX

press.princeton.edu



ISBN: 9780691203928
ISBN (pbk): 9780691206417
ISBN (ebook): 9780691249537

British Library Cataloging-in-Publication Data is available

Editorial: Abigail Johnson
Production Editorial: Terri O'Prey
Text Design: Wanda España
Cover Design: Wanda España
Cover image by Jake Postiglione
Production: Jacqueline Poirier
Publicity: William Pagdatoon

This book has been composed in Arnopro and ProximaNova

Printed on acid-free paper.

Printed in the United States of America

10 9 8 7 6 5 4 3 2 1

*To all my students, past and future.*
*You inspire and motivate me every day.*

*Un giocatore lo vedi dal coraggio,*
*Dall'altruismo e dalla fantasia. . .*

*(You'll see a great player from their courage,*
*their generosity and their creativity. . .)*

F. De Gregori, *La leva calcistica della classe '68*

# Contents

# Preface

Like many authors, I am sure, I never really set out to write a textbook, and I stumbled into this adventure thanks to a chain of serendipitous events. A few years ago, I started noticing a widening gap between the skills traditionally taught in physics and astronomy programs, and the ones that seemed to be most effective both in academic research and in the nonacademic job market. Industry employers and the world of research alike were entering a data-intensive regime, where the ability to extract meaningful insights from data seemed to be of paramount importance; at the same time, data were becoming bigger, richer, and more complex, and getting them to tell a story required increasingly specialized tools. Ignoring this gap made little sense to me. When I helped set up a brand new major at the City University of New York (CUNY), called "Applied Computational Physics," I proposed to create a new course, "Machine Learning for Physics and Astronomy," to be included in our required curriculum. My crusade was well received, leaving me, about six months later, with a course to teach in the major's first semester, an enthusiastic first cohort, two draft Jupyter notebooks, and a handful of good ideas.

The dramatic rush that ensued—whose true scope I would only fully understand en route, as many first-time instructors will tell you—was one of the craziest times of my life as a faculty member. I had, of course, many excellent books to look at—Jake VanderPlas' *Python Data Science Handbook* in primis—but I wanted to provide students with materials that were: 1. Accessible to undergraduate students and beginning practitioners; 2. Focused on concepts and with little mathematics, both because I didn't think that it was necessary and because I needed to serve students from different majors; 3. Relevant for research cases and not excessively polished, like many of the "classic" introductory data sets in computer science; and 4. Meant for practical purposes, inclusive of notebooks and exercises.

The result of many iterations over that initial madness, which took the form of teaching a similar course to several cohorts of undergraduate and graduate students, together with various summer schools and other informal classes, is distilled in this textbook.

Of course, writing a real book was a challenge in itself. I thought that I was doing a decent job at teaching the materials, but the idea of crystallizing my thoughts in a book was crazily intimidating. People who wrote similar books were true experts and stellar researchers. I felt that I was not well prepared enough, that if I couldn't write a great (or even—gasp!—*the best*) book, there was no point: Others would always be able to do it better. Then, at some point, I realized that this hardly mattered. Could I write a *useful* book? Was I willing to put in the time and the effort to face a blank (Overleaf) page, day after day until a useful book was born? Suddenly, this seemed a lot more manageable. I didn't need to be the best person to write this book. But I could show up and do the best that I could. And so that's what I did. I kept writing, and studied, and read papers, and wrote more, and then I asked for feedback, and I continued, until I was done.

As far as I can tell, I have reached my goal: This is definitely not the best book, but I think it can be a useful book. A book that a beginning practitioner can read and find approachable, a colleague who needs to incorporate machine learning in their class can use without a steep learning curve, a researcher who is exploring new ideas can pick up with ease, a self-learner who wants to read something accessible and relatively math-free can enjoy. My hope is that anyone who understands linear algebra and has some free time can read and learn from this book.

The aspect of machine learning that I am most excited about concerns enhancing scientific creativity. The more mechanisms we have to formulate questions and look for answers in original ways, the better scientists we can be. This explains my unusual choice of epigraph—what does an inspired soccer player have to do with a scientist using machine learning? My answer is that they have a lot in common: Science, like soccer, is at its best when it's daring, creative, and collaborative. With this book, I hope to give a tiny contribution to making machine learning tools more accessible to more people and to allowing them to be more creative in their scientific endeavors.

The unanticipated consequences of writing this book are also the best ones: One, I learned a lot about applications of machine learning, and two, I got to know and interacted with many great people. Going from the vaguely formed idea of writing a book and having some notes to the finished product was a long and winding road, and I would have fallen off the track multiple times if I hadn't had so much support.

My first big thank you goes to all the students who bore through my ML classes over the years. Without you, this book wouldn't exist (we can argue about whether that would be the better scenario). I am especially grateful to students from my first cohort—in particular, Harpreet Gaur, Hashir Qureshi, George Nwanwko, Charlie Meyers, and Kayla Ford. You kept me grounded and were brave enough to let me know that my assignments were a little over the top. Faraz Chahili also deserves a mention here, as an honorary group and class member, and a sufferer through my notebooks, as do Jake Postiglione and Olga Privman, shining stars of the Javioli team.

I am thankful to Ashwin Satyanarayana, who co-taught my first ML class with me and was crazy enough to embark in this adventure when I had little more than a syllabus and a few notebooks in hand. I learned so much about clustering from your crystal-clear lectures—any confusion you see here is exclusively my doing.

My long-term collaborators Eric Gawiser, Kartheik Iyer, Chris Lovell, Emille Ishida, and Andy Lawler taught me a lot about data analysis and machine learning in many conversations over the years. I hope to continue to rely on your knowledge and insights whenever I need help, which is very often.

I am deeply grateful to David Spergel, who encouraged me to teach an ML class at the Flatiron Institute and was one of the first catalysts of this endeavor. Our casual lunch conversation a couple of weeks before the beginning of my sabbatical culminated in me sending out the book proposal just a day before moving overseas and two hours after getting my wisdom teeth out. Without his encouragement and support (and the post-surgery drugs?), it's safe to say that I would have never dared to send the proposal out.

David Hogg was incredibly generous with his time and knowledge and met with me every week for an entire semester, just to help me put my thoughts in order. You are such a clear thinker, and I learned a lot (and had a lot of fun) during these meetings—I still miss them!

Some lovely colleagues were kind enough to provide comments on early versions of the draft, in particular, Benne Holwerda and Marcel Haas. I am forever grateful to both of you for your thoughtful feedback and mindful encouragement throughout the process. Benne actually deserves to be mentioned twice, as he was also one of the official book reviewers, together with Tomaso Dorigo, David Rousseau, and one more reviewer who chose to remain anonymous. I am very indebted to all of you for a great set of comments. The idea of adding a final chapter, which I hope will be a useful compendium of the whole book, comes from you, together with many smaller improvements. And even if I didn't implement all of your suggestions, for various reasons, please know that I have read and considered each one of them, and I am appreciative of your time and effort.

Many people contributed data sets or data processing ideas. I am thankful in particular to Sascha Caron, Giovanni Ossola, Steve Bickerton, Farnik Nikakhtar, Serena Di Pede, and Siddharth Mishra-Sharma, who helped me find good problems and lent their subject matter expertise when I felt out of my comfort zone. If you see anything poorly written or incorrect, the fault is all mine.

I had many casual conversations, in person, on Zoom, and on social media, with colleagues and friends who helped me with comments on the book, the notebooks, or the process of book writing: Chuck Keeton, Vishal Verma, Joel Zinn, Desika Narayanan, John Wu, Francesco Orabona, Saurabh Jha, Johannes Buchner, Giacomo Vianello, Austen Groener, and Roberto Trotta. Thanks for lending your time and expertise to my often poorly worded questions. A special thanks also goes to Alberto Bolatto; he might not even remember it, but his joke about

Inigo Montoya still makes me chuckle every time I talk about natural language processing.

I am grateful to Licia Verde for her thoughtful mentoring over the years and for hosting me during my sabbatical at the Institut de Ciències del Cosmos, Barcelona, where this project started, and to Julianne Dalcanton and the Center for Computational Astrophysics of the Flatiron Institute, New York, where this project was completed. I truly admire their talent as scientists and science communicators, and I am so glad to have received their support.

My editors, Jessica Yao and Abigail Johnson, have been an absolute joy to work with, and they have stayed encouraging, positive, and patient throughout this very long process. I am also deeply grateful to my production editor, Terri O'Prey, my copyeditor, Cyd Westmoreland, my illustration manager, Dimitri Karetnikov, and the rest of the team at Princeton University Press. They supported me with endless patience and clarity, and refrained from sarcasm as I tamed my addiction to excessive emphasis, I reluctantly accepted to leave LaTeX behind, and I learned the difference between vector and raster formats.

The strong ladies from my "EoX" peer support group have been an incredible resource during the past 10 years, and I am just so lucky to have them in my life. Thank you for everything.

Finally, a huge thank you to my amazing husband, Samir, and my lovely daughter, Clara. You have, respectively, tirelessly supported and joyfully sabotaged this project for the past three years, and I am the luckiest person in learning from both of you, every day.

Viviana Acquaviva
June 2022

# MACHINE LEARNING FOR PHYSICS AND ASTRONOMY

CHAPTER ONE

# Introduction to Machine Learning Methods

*I have a great supervised machine learning joke. . .*
*But you need to have heard a similar one before.*

## 1.1 WHAT IS MACHINE LEARNING?

The first question I am always asked when I see distant relatives at holiday gatherings is: What are you working on? And if for many years, "Astrophysics" was a popular yet often misleading answer (as I had to explain that I was not training to be an astronaut, and in fact, sadly, I would probably never even discover a new world), now that I've thrown machine learning into the mix, my answers have become even more murky and vague. However, I think it's important that we can explain without any jargon what we do, and as a consequence, I've spent many hours thinking about how to describe machine learning.

To the best of my knowledge/ability to explain, I would say that it's *the process of teaching a machine to make informed, data-driven decisions*. Examples of such decisions include recognizing and characterizing objects based on similarities or differences, detecting patterns, and distinguishing signal from noise.

In many ways, the boundaries and definition of machine learning as a discipline are fluid, and so far, it's often been approached without a scientific spirit of inquiry. But it is my opinion that this process should be subject to the same level of rigor and testing that any scientific investigation needs to endure. In this book, we will explore many ways to build, test, understand, and break machine learning models.

## 1.2 WHAT CAN WE DO WITH IT?

Once we've established a working definition of machine learning, a far more interesting question to ask is: What can we do with machine learning tools? And this is where the conversation can become really long. I won't strive for completeness here,

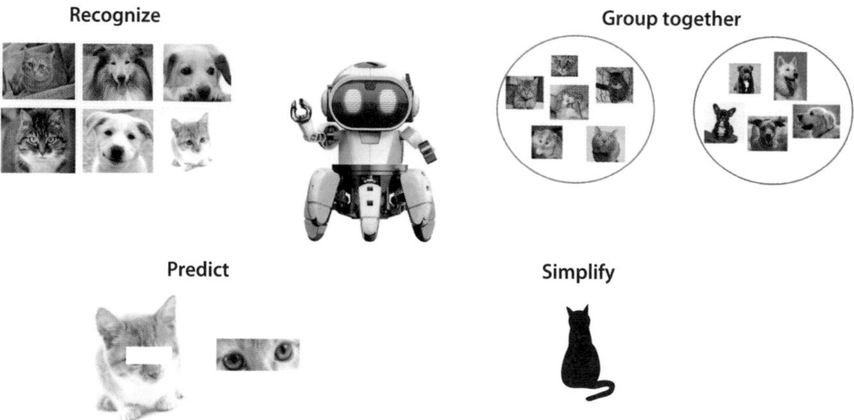


Figure 1.1: **A cartoon example showing some types of problems that machine learning can help solve.**

and there are many real-life applications that involve machine learning methods but are so complicated that they can't quite fit in a mold, for example, self-driving cars or AI (artificial intelligence) systems playing video games. But when reduced to building blocks of larger operations, I would say that often machine learning tools are used to do one of the following four things (see Figure 1.1):

- **Recognize** an instance of a certain type. For example, this could mean correctly labeling different type of animals, like cats or dogs, from images, or recognizing a specific person in a picture, as is done in social media "tagging" of people.
- **Predict** some property or information on the basis of some other information. For example, we may attempt to predict future behavior based on past behavior; use previous utility bills to predict the next one, or reconstruct a missing part of an image based on other examples of similar images.
- **Group together** objects that are similar, which can also be used to single out objects that are different ("outliers" in scientific parlance). For example, we might try figuring out how many different types of galaxies there are in a picture of the sky, or which strange-looking astrophysical sources are more likely to be artifacts of the camera. This type of application is referred to as *clustering*.
- **Simplify** the information contained in a complicated data set to condense it to its quintessential nature. For example, we could ask what's the simplest way to draw a cat so that it's still recognizable as a cat. This process has the double advantage of reducing the volume of data we need to deal with (which is desirable for manipulation, storage, and visualization purposes) and helping us understand what the essential properties are of a given category (in our case, cats). These are often called *dimensionality reduction* techniques by the initiated.

## 1.3 THE LANGUAGE OF MACHINE LEARNING

Unlike in the Fight Club,[1] the first rule of machine learning is to learn how to talk about it: Laying down very clearly what it is that we know and what it is that we want to know is incredibly important.

To begin with, the elements of a data set are often called *instances*; other commonly used names are "samples" or "examples." In the physical sciences, it is not uncommon to hear "observations" in reference to instances, but we will try to stay away from this confusing habit. For each one of these instances, there will be some properties that are known; these are usually things that can be measured, observed through experiment, or simulated. They are called *features* (or less technically, "input"). For example, if you are studying Newton's law of gravity, you might measure the distance and time of falling objects, and you might ask yourself about the final speed. In this case, the features of your problem are two: distance and time. If you are studying galaxies and you are observing them in different regions of the electromagnetic spectrum (say, for example, ultraviolet, visible, infrared, and radio), and you have one data point for each of those measurements, those will be your four features. Note that the features don't need to be numerical! They can also be of *categorical* type, like yes/no, or 0/1, or low/medium/high, or even have a descriptive quality like red/blue/green. Usually, it is necessary to map such types to numerical values before plugging them into ML machinery.

It is customary to organize a data set in rows and columns, where each row describes one instance, and each column contains the measured value of one feature associated with that instance. Therefore, usually the total size of your data set will be given by the number of instances times the number of features.

At times, there will be one or more additional pieces of information (properties; in the example above, the final speed) that you would like to estimate or predict, given the value of the features. This property is called *target*, or sometimes *label*, or more generally, output. In other cases, the output might not be a specific quantity, but rather a pattern, or a rule.

A summary plot of this basic terminology is shown in Figure 1.2.

Usually, the goal of an ML task is to build a relationship between input and output, which will then be described by our *machine learning model*. A model is a mathematical object that allows us to go from the input space of features to the output space of targets. Related terms include the words "method" and (more commonly) "algorithm," which usually refer to particular classes of mathematical relationships that can be used to build models. Examples of ML algorithms that you might have heard of include Random Forests and neural networks. They exist independently of our input and output, and once they are used for a specific problem, they define some rules on how we can build models.

---

1 https://en.wikipedia.org/wiki/Fight_Club

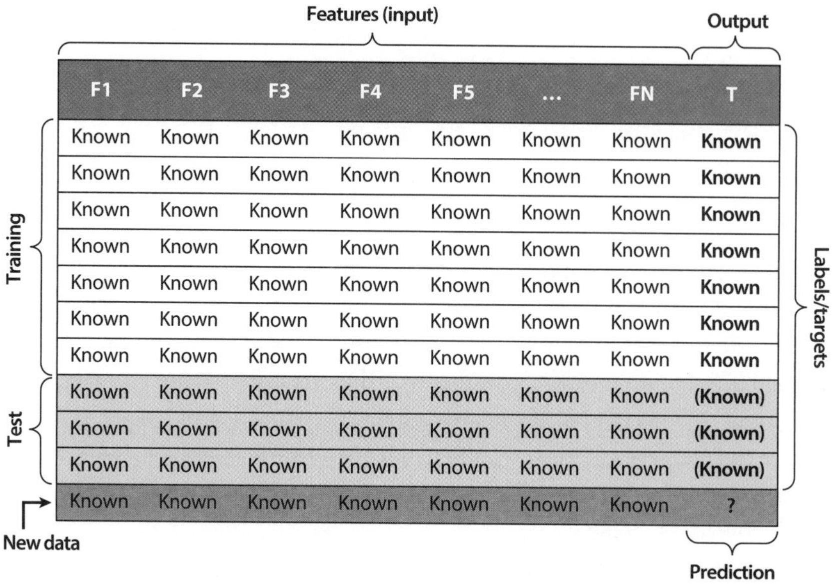

Features (input) | Output

| | F1 | F2 | F3 | F4 | F5 | ... | FN | T | |
|---|---|---|---|---|---|---|---|---|---|
| Training | Known | Known | Known | Known | Known | Known | Known | **Known** | Labels/targets |
| | Known | Known | Known | Known | Known | Known | Known | **Known** | |
| | Known | Known | Known | Known | Known | Known | Known | **Known** | |
| | Known | Known | Known | Known | Known | Known | Known | **Known** | |
| | Known | Known | Known | Known | Known | Known | Known | **Known** | |
| | Known | Known | Known | Known | Known | Known | Known | **Known** | |
| | Known | Known | Known | Known | Known | Known | Known | **Known** | |
| Test | Known | Known | Known | Known | Known | Known | Known | **(Known)** | |
| | Known | Known | Known | Known | Known | Known | Known | **(Known)** | |
| | Known | Known | Known | Known | Known | Known | Known | **(Known)** | |
| New data | Known | Known | Known | Known | Known | Known | Known | ? | |

Prediction

**Figure 1.2: A visual summary of some terminology frequently used in machine learning.**

One important distinction among machine learning methods is between *supervised* and *unsupervised* methods. The boundary between them is sometimes blurry, and their union doesn't cover the whole range of possibilities; some practitioners prefer to use different terminology or to do away with these denominations. Nonetheless, understanding the distinction between the two methods is, in my opinion, important, and we will discuss it below.

## 1.4 SUPERVISED LEARNING

In a supervised learning task, we assume that there is a collection of instances for which the target property is known (besides the features, which are assumed to be known for all data). This collection is called the *learning* set. For example, let's assume that each of the data points in the left panel of Figure 1.3 has a color associated with it: blue or green. We would like to learn to predict the color, based on each point's coordinates; gray indicates that the color is not yet known. This data set has two features (the x and y coordinates). Supervised learning consists of *learning by example:* we need to be shown some instances for which the color is known, in order to develop an intuition—in ML parlance, to build a model—of the relationship between coordinate and color. The learning set for this problem is shown in the right panel of the figure, and it contains 30 instances (the number of colored points).

To solve this problem, an ML algorithm will attempt to use the instances in the learning set to infer the rule that connects the coordinates to the color. If it is successful, when presented with another point for which only the features

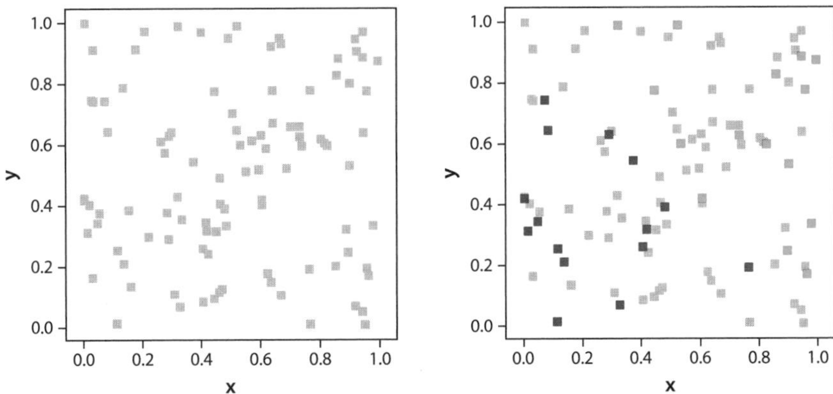


Figure 1.3: **A simple example of a supervised problem. We would like to learn to predict the color of a point, given its coordinates (points for which the color is unknown are gray). To do so, we are provided with a *learning set* (right): a subset of instances for which both the features (coordinates) and the target property (color) are known.**

(coordinates) are known (e.g., any of the gray-colored points in the figure), the algorithm will be able to make a correct prediction of the target property (color).

As a general rule, a *supervised learning method is only as good as its learning set* or at least it starts out like that. A training set that contains too few examples would not allow an ML model to learn the relationship correctly. Let's look at this simple example:

| input | output |
|---|---|
| 1 | 3 |
| 2 | 3 |
| 3 | ? |

Here I am providing two examples in the learning set, and I am asking you to predict the output of a third example. Most likely, you would predict "3" as the output, but alas, this is not the correct answer. We can try to improve by adding a few more examples to the learning set:

| input | output |
|---|---|
| 1 | 3 |
| 2 | 3 |
| 3 | 5 |
| 4 | 4 |
| 5 | ? |

Now, it is possible that some of you have already figured out what the rule is. But for the vast majority of us who still wouldn't know, let me rewrite the information above in a slightly more helpful form:

| input | output |
|---|---|
| one | 3 |
| two | 3 |
| three | 5 |
| four | 4 |
| five | ? |

I won't spoil the fun right here, but you can check your intuition (or give up and move on) by reading the note at the bottom of the page.[2] Hopefully, this toy example serves to show some important properties of a good learning set. The first is that it needs to be large enough so that our algorithms can figure out what the rule is. The second is that the choice of features matters; some representations of the data work better than others, even if they are based on the same information. In our simple case, deciding to spell out the numbers instead of using their mathematical representation served to emphasize which aspect of the data was important. This is relevant, because it is usually the job of the scientist to decide how to organize the information when building a data set.

### 1.4.1 Train and test sets

One important consideration in supervised learning is that it would not be wise to use all the objects in the learning set to infer the relationship between features and target (again in ML parlance, to train the model). This is best understood, in my opinion, by thinking about the scientific method itself. The process of training the model is akin to formulating a hypothesis. The next step then is to make predictions that result from that hypothesis, and to test them to verify whether they are correct. It should be clear that we cannot verify the predictions of the model on the same instances that were used in the training process, because our verification process needs to be independent of the training process. Another way to think about this issue is that we are interested in assessing how well our model can predict the target property of *new* data; objects that participated in model building are not new to the model.

Therefore, it is customary to set aside a subset of the learning set that does not participate in the model building. Once the training process is complete, we can use the model to make predictions for the target property of those objects and verify that they are correct. Or more generally, we can check how the model performs on that subset and decide whether we are satisfied or require further improvement. The subdivisions of the learning set used for training and testing a model are called—you guessed it—training (or train) set and test set, respectively. The idea is illustrated in

2 The model should return the number of letters in the input.

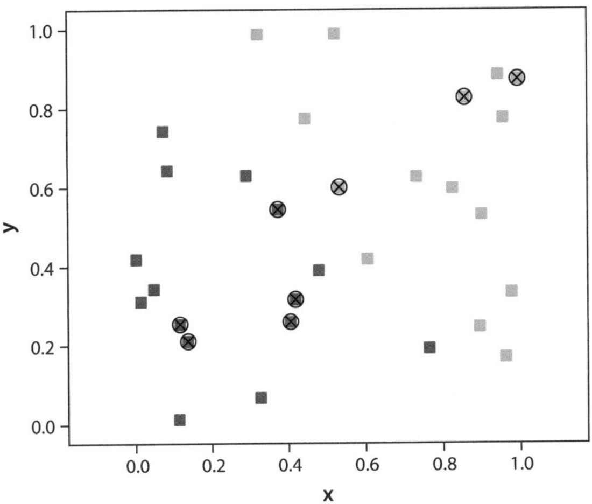


**Figure 1.4: The learning set (the set of objects for which the labels are known) from the right panel of Figure 1.3 is split in a training set (square points) and test set (crossed-out points). The test set does not participate in the training process. The performance of the algorithm on new data can be estimated by applying the trained model to the test set features to generate predicted labels (colors) and comparing them to the true ones.**

Figure 1.4. In the next few chapters, we will discuss at length what the best way is to split a data set into training and test sets.

The performance of a model can be expressed as rate of failure (error) or rate of success (score); these are just two equivalent ways of reporting how well a model works. We will often distinguish between the performance of a model when applied to the training set (the *training error* or *training score*) and the performance of a model when applied to the test set (the *test error* or *test score*). Finally, the performance of a model on *new* data, which were not part of our learning set, is called the *generalization error* (or *generalization score*) of the model. We don't have a way of calculating the generalization error directly, because the ground truth labels are not known. Therefore, we use the test error as a proxy for the generalization error.

Let's continue our preliminary investigation by looking at one more simple example. Imagine that we want to apply an ML method to the population of points from Figure 1.3. What would happen if our learning set was only made up of the blue objects, as in Figure 1.5? No matter how we decide to split in train and test sets, our model would learn that any combination of coordinates leads to predicting the color as blue, because it has never been shown that green points exist. And this is not even the worst problem. The problem is that if we then proceeded, like good ML practitioners, to verify our predictions on the test set, our constant prediction of "blue" would be correct 100% of the time. So we would get a false sense of confidence that our model is reliable, when instead it would fail miserably on the vast majority of the other data points in Figure 1.5, which (as we know from the previous sections) are mostly green.

This simple example illustrates one of the cruxes of machine learning techniques: Because they are driven by the data as opposed to relying on physical intuition, we are bound to make a fool of ourselves if we don't understand the data

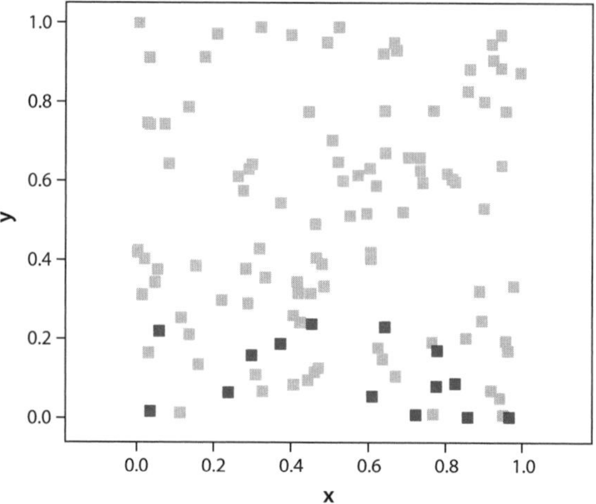


**Figure 1.5: The subsample of blue points is a dangerous choice of learning set, if our goal is to learn how to predict color, based on coordinates, for all the points in this diagram (same population as Figure 1.3).**

well. How can we save ourselves from this sorry destiny? In the case considered here, we should have noticed right away that the feature space (i.e., the range of x and y coordinates) spanned by our learning set is very different from the feature space spanned by the population we want to apply our model to (i.e., the gray points in Figure 1.5). This should have made us suspicious, because in general a learning set should be representative of the application domain, meaning that the learning set and application set should be statistically similar. If they are not, then we can hold no hope that the test error will be a good proxy for the generalization error. This is an important minimum requirement to add to the desired characteristics of learning sets: We'd like them to be large enough, representative of our application domain, and ideally organized in a way that optimizes the information content for the task at hand.

It's not always easy to make sure that these conditions hold, as we will see as we continue our ML journey.

### 1.4.2 Classification vs Regression

Another important distinction (although possibly more in terms of nomenclature than methods) in the realm of supervised problems is between classification and regression tasks. They only differ in the "output" part: in *classification*, the target property belongs to a discrete set of possibilities (in other words, a class). A simple toy example could be to correctly identify fruit in a bag on the basis of properties that can be measured through touch, such as height, width, weight, and shape. For something more relevant to daily life, recognizing people in pictures (the "tagging" of social media) is an example of a classification problem: The output can only be a specific person, and there is no notion of adjacency or continuity between outputs. As a result, classification algorithms will output a response that is either 100%

correct or 100% wrong. Either they recognize you (yay!), or they will mistake you for someone else (nay). When we evaluate the performance of a classifier, we will "count" (in more or less smart ways) the number of correct answers.

The other option is *regression*. In regression tasks, the output is a continuous variable (typically, a real number). For example, if we were trying to emulate Galileo and learn to predict the timings of different falling objects (without knowing the formula!) on the basis of distance traveled, our output would be time, probably in seconds. The output can take any value; in fact, the number of significant digits is only dictated by the precision of our measurements. As a consequence, in regression tasks, we cannot evaluate our model by asking for an *exact* answer; instead, we will assess how close we are to the correct value. Unlike in classification problems, where we are either correct or wrong, if the correct answer is 3.1415, predicting 3.0 and 10.0 are very different results.

It is important to note that the distinction between classification and regression problems can be quite fuzzy. For example, many problems in which the output is a decision ("Will it rain later today?" or "Will I be approved for a loan?") rely, both from a conceptual and a mathematical perspective, on the idea of a threshold. Even if the final output is discrete, the features will be implicitly mapped to a probability, which is a continuous variable, and then a threshold will be chosen to separate the classes (e.g., if the probability of rain is assessed to be more than 50%, the answer will be "yes"). Therefore, while these types of problems technically meet the definition of classification, they can be easily recast or thought of as regression problems. In fact, casting them as such might be advantageous, because outputting a probability retains more information about how confident we are in our prediction. Our model might indicate "no rain later" for two different sets of conditions, but if all weather models indicate a high pressure front, there is not a single cloud in the sky, and you are in the desert, the prediction for no rain will be much more solid than in a scenario of high humidity, low pressure, and a cloudy sky.

However, there are some classification problems that are truly "discretized" in nature. For example, one of the classic data sets for machine learning applications is the digits data set [Le Cun et al., 1989], where the machine is tasked with "reading" images of hand-written digits. The output is one of ten classes, the numbers from 0 to 9. In this case, the classes are truly separate from one another: 0 is not more similar to 1 than it is to 8 or 9. Another, more modern, example is the CIFAR-10 data set [Krizhevsky and Hinton, 2009], which is an image recognition problem with ten possible outputs, including birds, cars, and airplanes, as seen in Figure 1.6. Again, there is no obvious mapping of the distance between classes. This lack of contiguity among classes is what characterizes pure classification problems.

Finally, let me state again that the difference between classification and regression tasks is unrelated to the features of the problem and is only determined by the

**Figure 1.6: The CIFAR-10 data set is composed of tiny, and therefore blurry, images belonging to 10 distinct classes (for example, cats, automobiles, and birds). Picture from the PyTorch [Paszke et al., 2019] website. Copyright ©2013 Valay Shah**

target property, or output. Somehow, this always seems to be a tricky point for my students, so I just state it again here for emphasis.

## 1.5 UNSUPERVISED LEARNING

In unsupervised learning tasks, there are no labeled examples. Rather than predicting a specific quantity or property from the features, we are trying to discover patterns in the data. In a way, the target of an unsupervised learning algorithm is a pattern, as opposed to an unknown property.

### 1.5.1 Clustering

A vast subset of unsupervised learning tasks has to do with counting or grouping objects in a smart way; these applications are known as *clustering*. For example, imagine that you want to count how many friends appear in your childhood (or Facebook) photos. A successful clustering algorithm would be able to group

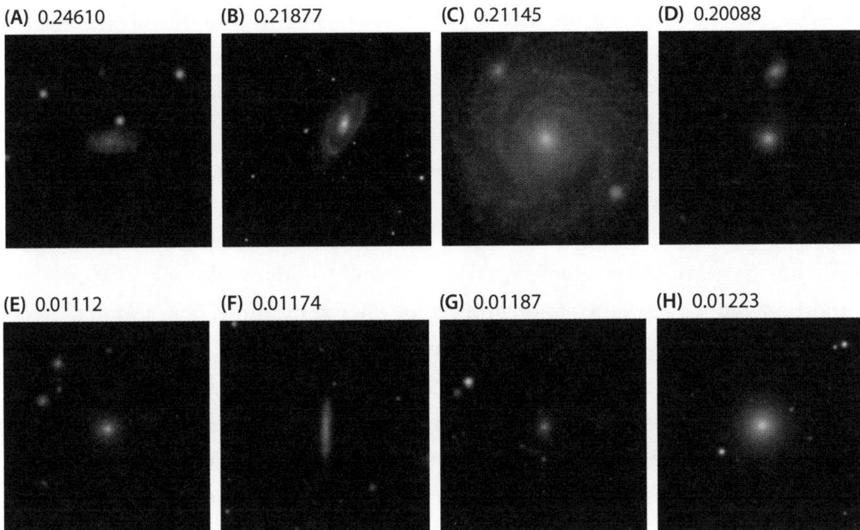


**Figure 1.7: A sample set of galaxies observed by the Sloan Digital Sky Surveys (SDSS), showing different types of galaxy morphologies. Clustering could be used to determine from data how many fundamental morphological types exist. Reproduced with permission from [Dieleman et al., 2015].**

together all instances of the same person, which would become a cluster; you could then count the number of occurrences. This happens without needing to know anything about the identity of the individuals beforehand; in other words, without the supervision process. Another application that we all, sadly, have become familiar with has been visualizing outbreaks of a disease; most recently, COVID-19. The outbreaks are usually represented with noncontiguous circles on a map. The clustering process can be used to determine the location (centers) and size (radii) of those circles that best represent the data.

Unsupervised learning methods often can be used to check a scientist's intuition. One historic example is the Hubble classification of galaxies. Edwin Hubble was one of the pioneers of extragalactic astronomy, and he proposed to divide galaxies into the main categories of elliptical, spiral, and irregular galaxies, according to their morphology (shape), as illustrated in Figure 1.7. There were further subdivisions, but let's ignore them for the time being. His reasoning was motivated by what he (largely incorrectly) believed to be the evolutionary track of galaxies. Nowadays, the task of automated morphology classification has become quite crucial, because we have observed many millions of galaxies, and upcoming instruments, such as the Vera Rubin Observatory or the James Webb Space Telescope are set to observe many more; with more than 200 billion of them in the Universe, we won't be done any time soon. This problem can be solved in a supervised manner, by deciding which classes to use beforehand (e.g., see [Dieleman et al., 2015]); in the case of Hubble's original proposal, they would be elliptical/spiral/irregular. Humans then would need to build a learning set by providing visual classifications for an appropriate number and range of galaxies. However, it might be more advantageous to cast this problem as an unsupervised task, as in [Hocking et al., 2018]. In this case, our ML model will decide for itself which classes to pick by selecting its own criterion for forming clusters, which will determine which objects are assigned to each cluster, as well as (possibly) how many clusters will be found. This process creates a completely data-driven classification scheme, which might differ significantly from the one proposed in a supervised scheme and could reveal something new and important about galaxy formation and evolution. For example, if we found a "new" type of galaxy that is different enough from others to warrant its own cluster, this discovery might point to a different evolutionary pattern for those galaxies. Another advantage of the unsupervised approach is that no learning set is needed. However, the classification (or regression) schemes provided by unsupervised models are often harder to interpret, as it might not be easy to understand what classification criterion has been chosen (i.e., what the common properties of objects in the same cluster are), so human intervention might still be required. Sometimes, these two-step approaches are referred to as *semi-supervised learning*.

### 1.5.2 Dimensionality reduction

Another broad subcategory of unsupervised learning tasks refers to "simplification" processes, known as *dimensionality reduction* (DR). These techniques are typically used to make a data set smaller, and therefore more manageable and easier to visualize, with minimal loss of information. Their eventual success relies on the amount of redundancy present in the original data set and on the efficiency of the chosen compression technique.

DR methods tend to work in one of three ways. The simplest approach consists of retaining the original features and just selecting those that are expected to be more meaningful. The second approach is to remap the feature space to a different one, in which fewer components can express the highest amount of variance of the original data. Finally, a third approach consists of learning a *manifold*, a nonlinear space in which some interesting properties of the original data set (e.g., pairwise distances or dot products between elements) are preserved. See [Sorzano et al., 2014] for a review. Often, DR techniques can be used as a preprocessing step for clustering algorithms.

A simple graphical representation of clustering and dimensionality reduction is shown in Figure 1.8.

The most fun, useful, and creative applications of machine learning happen when we learn to mix and match all the techniques. For example, we can decide to learn (unsupervised) feature representation as a way of preprocessing a labeled data set, obtaining a more efficient and helpful representation of a learning set, and

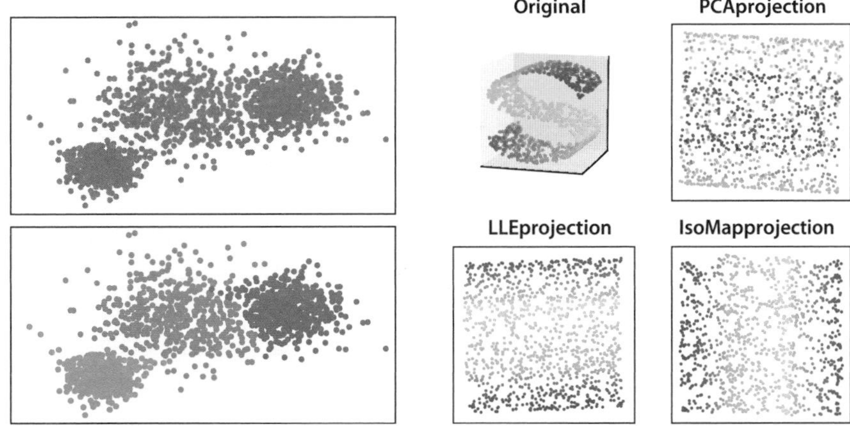


Figure 1.8: **Left: An example of a clustering algorithm applied to a set of points. The algorithm assigns each point to one of three clusters. Right: Different examples of dimensionality reduction algorithms, with varying degrees of information loss. Figure from the `sklearn` manual [Pedregosa et al., 2011].**

then apply a supervised learning method. Or we can recycle pieces of pre-trained algorithms to accelerate the training and learning process for a different but related data set or problem. Science progresses with rigor and creativity; machine learning methods provide a versatile range of techniques to tackle problems in innovative ways.

## 1.6 MACHINE LEARNING VERSUS INFERENCE

Machine learning methods can be used to solve many problems, as we will see as we continue our journey. But of course many research questions would be better approached from a different perspective, or put more simply, other approaches can be equally beneficial. If we think of machine learning as a means to build an implicit relationship between input and output (whether the output is a quantity, a rule, or a pattern), the alternative is often seen as classical *inference*, where we explicitly specify what the output would look like *as a function of the input features, with some parameters chosen by us*. This model could be a simple mathematical formula or the result of a complex numerical simulation; the important aspect is that we are able to predict y, the output, for a given value of x and the model parameters. The goal of the inference process is to determine the model's *parameters*, unlike in machine learning, where the goal is (usually) to make a prediction.

Let us consider a very simple example.

Suppose that a friend comes to you and says: I have this list of time measurements of a car moving at uniform speed. My wicked physics teacher asked me to figure out where the car would be at time $t = 12$ seconds. I have no idea of how to proceed. Can you help? These are my data:

| time (s) | distance (m) |
|---|---|
| 0 | 5.1 |
| 1 | 5.5 |
| 2 | 8.4 |
| 3 | 11.1 |
| 4 | 11.8 |
| 5 | 14.4 |
| 6 | 16.1 |
| 7 | 19.5 |
| 8 | 20.2 |
| 9 | 23.1 |

So you—being a good scientist—would certainly know that you can use this well-tested formula for the relationship between the input (time) and the output (distance traveled):

$$d = d_0 + v \times t \tag{1.1}$$

This model has two parameters: the initial distance (or coordinate), $d_0$, measured at time $t = 0$, and the speed, $v$, which is constant, as we know from your friend's description. But how can we work out what the correct values are to plug in the formula for $d_0$ and $v$? This process is called *parameter fitting*.

There are many great books that describe this process in detail (some of which you can find in the References at the end of this book), so I'll just mention a very basic approach.

First of all, since we have only two coordinates, it's always a good idea to look at our data. The left panel of Figure 1.9 shows that the data lie approximately on

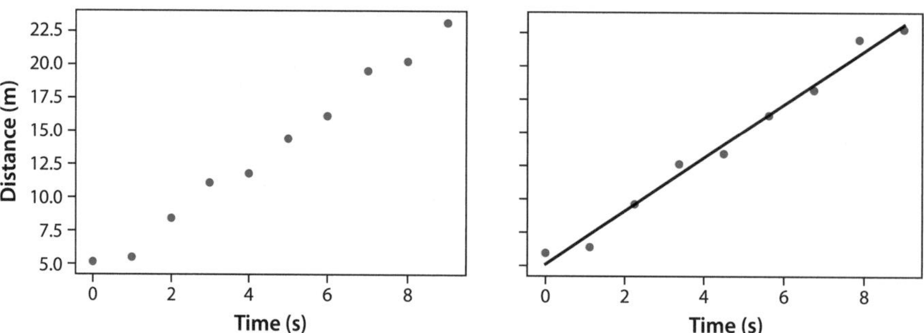


**Figure 1.9: A simple example of fitting data by using a linear model.**

a straight line, as expected; the deviation can be interpreted as the consequence of statistical error (measurement noise). The parameters $d_0$ and $v$ that we want to measure are the y intercept and the slope of this hypothetical line.

The main idea of parameter fitting is to try out many different combinations (in this case, pairs) of parameter values until we find something that works well. So we need at least three steps: 1. Decide on a range of possible values to try out; 2. Decide how to pick parameters; and 3. Establish an evaluation criterion that tells us whether a model is good or bad.

For the first step, a visual inspection of our graph tells us that probably the y intercept of a line that goes approximately through the points will be between 4 and 6, and the slope of the line will safely be between 1 and 3 (we can get a quick estimate by calculating slopes as rise over run for a couple of time intervals). Because our data space is very small, we can use a brute-force approach, dividing our ranges in equally spaced intervals and trying out every possible combination. We can either decide the spacing between different values we want to consider, or the total number of points; for this example, let us consider 100 values, which corresponds to a 0.02 spacing. The third step requires building a measure of the distance between the predictions of each model (i.e., the 10 distances provided by the model, given the 10 times, slopes, and y intercepts) and the observed data (the nine measured distances). In the simplest case where there is no uncertainty associated with each measurement (which is deeply unphysical!), we can use the square of the Euclidean distance between the 10 distances predicted by the model ($D_i^m$) and the 10 observed distances ($D_i^o$):

$$\Sigma_{i=1}^{10}(D^m(d_0, v, x_i) - D_i^o)^2. \tag{1.2}$$

Better models should be closer to the observed data, so the combination of parameters that generates the minimum distance can be considered the winning model, or *best fit*. In the real world, where every measurement has an uncertainty associated to it, the distance should be weighed against the uncertainty, so that points with large measurement errors contribute less to the total distance. If some assumptions about the distribution of uncertainties can be made, this process corresponds to writing a $\chi^2$ distribution, or a *likelihood*. As already mentioned, there is a vast literature on this subject; see, for example, [Hogg et al., 2010] and [Hastie et al., 2001].

To summarize, we pick every possible pair of ($d_0$, $v$) according to the chosen range and spacing, calculate the above metric, and choose the pair that minimizes it. We can do this easily as a Python exercise. We start by choosing an appropriate range and spacing for the parameter values:

```
slopes = np.linspace(1,3,101)
intercepts = np.linspace(4,6,101)
```

Then we define our model and fitness metric, the sum of squared errors:

```
def model(x,m,b):
return m*x+b

def se(m,b,x,y):
return np.sum(((model(x,m,b) - y)**2))
```

Finally, we calculate the squared error for all combinations of parameters and choose the ones that lead to the minimum error:

```
square_errs = np.array([[se(m,b,x,y) for b in intercepts] for m in slopes])
indices = np.unravel_index(square_errs.argmin(), square_errs.shape)
bestm, bestb = slopes[indices[0]],intercepts[indices[1]]
```

Through this process, we end up with a minimum distance value $d_{\text{min}} \sim 3.65$, corresponding to a y intercept ($d_0$ in the language of our problem) of 4.34 and a slope ($v$) of 2.04. The corresponding line is plotted in Figure 1.9, so we can convince ourselves that our procedure works.

Finally, we can use the equation of the line we found—in other words, our model—to predict the coordinate of the car at other times, for example, the $t = 12$s required by the physics problem:

$$d = d_0 + v \times t = 4.34\,\text{m} + 2.04\,\text{m/s} \times 12\,\text{s} = 28.82\,\text{m}. \tag{1.3}$$

What would be the corresponding approach in machine learning? The main difference is that we won't explicitly write out a model, so we won't write parameters or a likelihood; in general, we are not required to be able to predict the observed outcome of a given combination of parameters. However, in practice the choice of ML algorithm or method that we make will affect what kind of input/output (I/O) relationship can be represented by our model, as well as its ability to learn from the training data.

In this case, we are dealing with a supervised learning problem, and we have 10 points in our learning set. As we know from the previous sections, we need to split them in a training set and a test set. For now, let us assume that we will use seven of them for training and three of them for testing; the test/train split can be selected randomly. Note that, in general, this is not allowed in time series problems; but for the purpose of our problem, the fact that our independent (input) variable is a time is not relevant, so making this choice is OK.

Our problem is a regression problem, because we are predicting a continuous variable. So we need to pick an algorithm that can handle regression, and the metric we use to estimate how good our model is will also be sensitive to the distance between predicted and observed points; for example, it could be the mean squared error (MSE), which is in fact the square of the Euclidean distance along the y axis, averaged over the number of points.

Even if we are not properly defining machine learning algorithms yet, for the sake of the argument, let us try out two very simple ones: a linear regressor and a decision tree. You can see the implementation in the lecture notebook "Straight Line with ML.ipynb."

We can select a training set for each model, which contains seven points from the learning set. In `scikit-learn` or `sklearn` [Pedregosa et al., 2011], the Python package for machine learning that will be our main software library throughout the book, we can do this easily by using the auxiliary function `train_test_split` and fixing the random seed for reproducibility:

```
np.random.seed(10)
X_train, X_test, y_train, y_test = train_test_split(x, y, test_size=3)
```

In this case, the training set is made by instances with time coordinates 6, 3, 1, 0, 7, 4, and 9 s.

After training each model, we can ask them to predict the y coordinates (i.e., traveled distances) of the points in the test set, which have x coordinates 8, 2, and 5, respectively. This is an example for the Decision Tree algorithm (note that `scikit-learn` is imported in code as `sklearn`, and for this reason, we will use the latter notation throughout the book):

```
from sklearn.tree import DecisionTreeRegressor
treemodel = DecisionTreeRegressor()
y_pred_tree = treemodel.fit(X_train.reshape(-1, 1),y_train).predict(X_test.reshape
(-1, 1))
```

The responses of the two algorithms are as follows:

| | x = 8 | x = 2 | x = 5 | MSE |
|---|---|---|---|---|
| True values y(x) | 20.2 | 8.4 | 14.4 | |
| Prediction LR | 20.9 | 8.42 | 14.7 | 0.18 |
| Prediction DT | 19.5 | 5.5 | 11.8 | 5.22 |

The built-in evaluation process of the machine learning approach tells us immediately that the linear regression model is superior to the decision tree model, because the mean squared error on the test set is lower; the decision tree model is quite severely underestimating each prediction.

Just as in the inference exercise above, we can now use either model to predict the traveled distance at time $t = 12$ s; the prediction of the linear regression model is 29.2 m, and the prediction of the decision tree model is 23.1 m.

### 1.6.1 Who fit it better?

The simple example above showed us two approaches to solving the same problem: one using classic inference, the other relying on machine learning methods. Both can be used to output a prediction. So which one is better? There is no hard-and-fast rule, but we can try to focus on some differences, and perhaps debunk some myths at the same time.

In the inference approach, we choose the functional form of the input/output relation explicitly, in parametric form, and we optimize the *model parameters*. This strategy is usually convenient when we have a good understanding (e.g., from physical principles, as in this case) of which variables matter and how different variables are related, or when we are looking at a very simple problem and data set. It is intuition driven; I like to joke that it is only as good as the scientist, but in general it's only as good as the model. If our choice of model is unphysical (e.g., if we had tried to model our data with a sinusoidal curve instead of a straight line), the parameter values we would have obtained would have not made any sense, and the model would have no predictive power.

Machine learning approaches are, in principle, more model agnostic, although as we saw above, the results obtained can be significantly algorithm dependent, because the choice of algorithm may shape the forms of input/output relations that we can explore. Even when we use very flexible ML methods, such as deep neural networks, which can alleviate this problem, they can only be good as the data that are used to build them. As a general and simplistic rule, I think that when we understand the physics but don't have data, we should use a probabilistic approach to model fitting, and when we don't understand the physics but we have data, we should use machine learning. Figure 1.10 summarizes some more of my thoughts on either approach; often, the synergy between these two approaches can be most powerful.

As a final note, I should add that distinguishing between probabilistic inference and the machine learning approach as I did here is too naive, and rather incorrect. There is more overlap between the two methods than Figure 1.10 implies, and ML methods can serve the purpose of probabilistic inference very well (e.g., see the recent review of simulation-based inference [Cranmer et al., 2020]). However, I think this framework is useful for building understanding of the scope and limits of ML methods for beginning practitioners, and I will stand by it for this introductory textbook.

### 1.6.2 The black box issue

Even after solving a problem using machine learning, the input/output relationship is never obtained explicitly, so the process of making predictions for new instances consists of feeding the trained algorithm new data and receiving some numbers

| Machine learning | Model fitting |
|---|---|
| • Data-driven (only as good as the data)<br>• Usually generalizes poorly (model derived using some data can't be applied blindly to different data)<br>• Interpretation is possible but might be nontrivial<br>• Fast(er)<br>• More robust/accommodating of mixed and missing data<br>• Allows serendipitous discoveries | • Intuition or model-driven (only as good as the scientist :) )<br>• Generalizes well if physics is well understood<br>• Easier to interpret<br>• Might be computationally intensive<br>• Dealing with heterogenous data often a pain in the neck<br>• Leads to loss of information if models are too simplistic |

**Figure 1.10: Some advantages and disadvantages of the two approaches. My claim is that synergy is often the best strategy. Figure from [Acquaviva, 2019].**

in return. The fact that we cannot simply write down an equation has gained ML models the infamous moniker of "black boxes." However, there are many ways of gathering insights on the nature of the trained model, and in my opinion, a good scientist would want to open the black box. This approach has the double advantage of validating the model and possibly gaining new information by analyzing the reasons for the algorithm's response. We will look at some ways to gather intelligence from ML models throughout this book.

### 1.6.3 There is no magic!

Although I do, in many ways, "believe the hype" about machine learning, and I think it is important for scientists to become proficient with these methods, I also think we should keep in mind that ML techniques are just tools of the trade, not the trade itself. When I was in graduate school, the most useful class I took was one about numerical methods; we learned from a giant book titled Numerical Recipes; I've often joked that ML methods are the Numerical Recipes[3] of this decade. In fact, many ML algorithms are nicely packaged, linear-algebra-based sequences of operations. They enable us to solve some problems in new, or more efficient, ways; as always, though, it's part of a scientist's job to understand how to tackle a difficult problem, with or without machine learning. One of my hopes for this book is to generate some understanding of when machine learning can help and of the strengths and weaknesses of specific methods. In all cases, and especially when we lose some transparency because of complex mathematics, it is essential to stick to the rigorous process of hypothesis testing spelled out by the scientific method.

---

3 http://numerical.recipes/

## 1.7 REVIEW AND DISCUSSION QUESTIONS

Note: Questions and exercises marked by ** are more complex, open-ended, or time consuming.

**Exercise 1.** For each one of the scenarios described here, answer the following questions. 1. Is it a supervised or unsupervised learning problem? 2. Is it a classification or regression task? 3. What could be useful features (data) to collect? You should be able to motivate your answers; several of them can be interpreted in multiple ways, so the argument will be more important than the answer!

Scenario 1. Email providers placing emails in the "Spam" folder.
Scenario 2. House selling prices: imagine that you are a real estate agent and want to come up with a data-driven listing price.
Scenario 3. Predicting used car prices (same as before, but now you sell cars).
Scenario 4. Twitter showing "trending topics."
Scenario 5. Recommending sizes for clothes bought online.
Scenario 6. Guessing a college student's major from personal information.
Scenario 7. Counting the number of people appearing in a collection of photos.
Scenario 8. Predicting the number of hours a college student will sleep tonight.

Note: This also works well as class exercise, or think/pair/share.

**Exercise 2.** Popular services like Netflix or Spotify include a "recommendation engine" for movies or music. Discuss similarities and differences in how they collect information, what information they collect, and how they use it to provide suggestions.

**Exercise 3.** For each of the following problems, discuss whether they could be partially or completely solved with machine learning. If the answer is positive, specify whether you would use supervised or unsupervised learning, the potential features, target properties if relevant, and potential process of data collection:

Problem 1. Determining the constant of gravity.
Problem 2. Determining the specific heat of a liquid.
Problem 3. Counting the number of sources in an astronomical image.

**Exercise 4.** ** Come up with a problem (related to your area of research or personal interest) that is better solved through classic inference and with a problem that is better solved through machine learning.

## 1.8 PROGRAMMING EXERCISES

**Exercise 1.** The linear model we used should be mathematically equivalent to the linear regression ML model, but the predictions for the distance at time $t = 12$ s are slightly different. Can you figure out why? Hint: There are (at least) two different reasons!

**Exercise 2.** ** Modify the "ModelingStraightLine.ipynb" notebook to include measurement errors for all the measured distances generated from a normal distribution with mean $= 0$ and variance $= 2.0$ (set the random seed to ensure that your results are reproducible). Assume that the errors on the time measurements are negligible. Modify the evaluation metric for the model to inverse-weigh the uncertainties. Does the best fit model change? Why or why not?

**Exercise 3.** ** Use the Dark Energy exercise notebook ("DarkEnergyFromSupernovae.ipynb") to find evidence for the existence of Dark Energy from supernovae data with classic inference methods.

CHAPTER TWO

# First Supervised Models: Neighbors and Trees

*Roses are red*
*Violets are blue*
*This is a binary classifier*
*The outputs are only two.*

## 2.1 BUILDING AN ML MODEL

It is now time to build our very first machine learning model! First of all, we need to understand what *building a model* means in practice. We will set the foundation now and will keep refining this sequence of operations as we progress on our ML journey. Items flagged with an asterisk are the ones that will change as we progress.

The fundamental steps of building a supervised model are the following:[1]

- Arrange the data into features and target arrays.
- Split them into a training set and a test set*.
- Select the ML algorithm you want to use, and its parameters*.
- Build the model by applying the selected algorithm to the training set, which will create a tentative (implicit) input-output relationship. If you are using `sklearn`, this corresponds to applying the "`.fit`" method.
- Apply the model to the features of the test data to predict the target property of the test data. If you are using `sklearn`, this corresponds to applying the "`.predict`" method.
- Estimate the performance of your model by using an appropriate evaluation metric to compare the predicted and true target property of the test data.
- Rejoice (unlikely), or figure out what is not working out and repeat (likely).

1 This list was loosely adapted from [VanderPlas, 2016].

Let us now build our first model, which will try to predict whether planets are habitable.

### 2.1.1 Problem: The search for habitable planets and the Planet Habitability Lab database

The search for intelligent life beyond Earth has deep roots in the history of humankind. For many scientists and nonscientists alike, looking out at the many stars of the night sky raises the question: "Are we alone?" One way to start looking for an answer is by searching for other places where life could have developed. In the past few decades, our hopes (or fears) of finding other forms of life have been refueled by the launch of several new instruments, such as the Kepler and K2 missions [Kepler] and the ongoing Transiting Exoplanet Survey Satellite TESS [TESS]. They have discovered numerous planets outside our Solar System, bringing the tally of confirmed exoplanets from just a handful to more than 5,000 within just over a decade;[2] some of them are shown in Figure 2.1. Of those planets, many have physical conditions that prevent the development of life; for example, they are too hot or too cold. Therefore, scientists are now focusing on finding *habitable* planets, generally defined as those whose density and temperature conditions are thought to be compatible with the development of life as we know it (and keeping in mind that we have one lonely example of how that happened). The catalog of confirmed exoplanets is publicly available (and constantly growing!), and it will be the basis for our first exploration of an ML model.

### 2.1.2 Data set and preliminary exploration

The actual data set available on the Planet Habitability Laboratory website[3] contains data for thousands of planets and collects a variety of features, but we will start our investigation with a small selection of samples and features. We will consider a learning set composed of 18 instances. For those planets, we have the information shown in Table 2.1.

The golden rule (made up by me) of data science is to *know your data* before starting out. This is easy to do in this case, because the data set is very small and can be examined by eye, but we will develop more formal techniques as we go along.

At the very least, we should do the following:

- Figure out the size (number of examples / number of features) of our data set.
- Check whether there are missing data, and decide how to handle them.

---

2 https://exoplanets.nasa.gov/alien-worlds/historic-timeline/

3 http://phl.upr.edu/projects/habitable-exoplanets-catalog

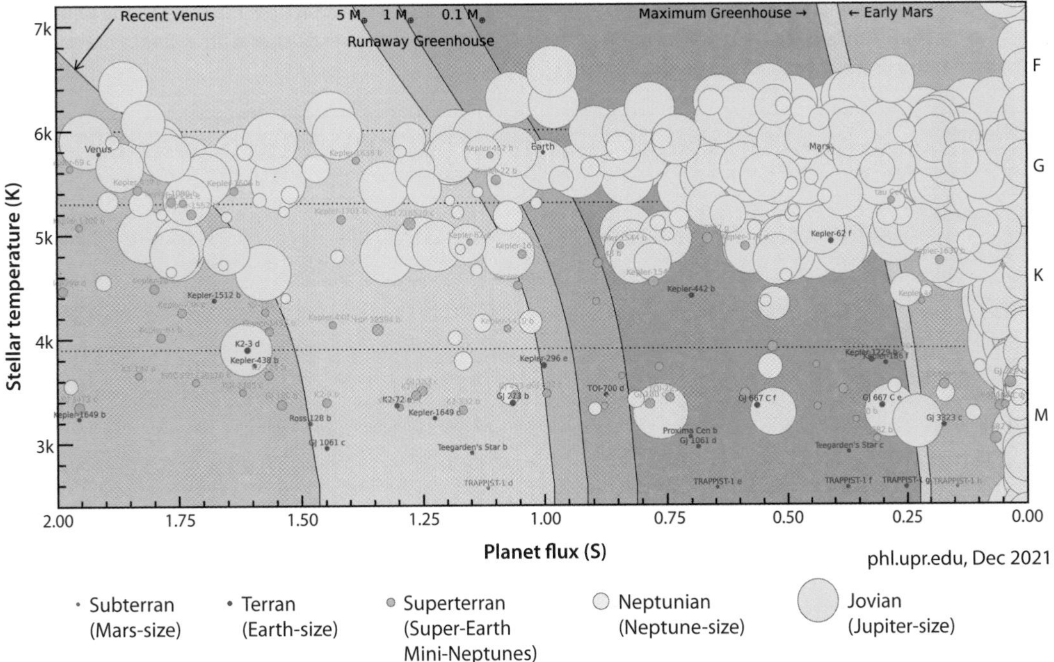


**Figure 2.1: The figure shows all planets near the habitable zone (the darker green shade is the conservative habitable zone, and the lighter green shade is the optimistic habitable zone). Only those planets less than 10 Earth masses or 2.5 Earth radii are labeled. The different limits of the habitable zone are described in [Kopparapu et al., 2014]. Size of the circles corresponds to the radius of the planets (estimated from a mass-radius relationship when not otherwise available). Credit: Planet Habitability Lab at University of Puerto Rico Arecibo, reproduced with permission.**

- Check whether all the features are in a similar numerical range (and if not, decide whether we need to normalize them), and determine whether there is something strikingly unusual about the distribution of their numerical values.
- In a classification problem such as this one, check whether the data set is very imbalanced (i.e., one or more classes are much more heavily populated than others).
- If possible, develop some intuition on how well we expect our model to work: Are these features meaningful? Do we have enough examples? I can't stress enough how many times I found a bug in my code because the performance was too good to be true, and similarly, many other times I've (nevertheless) persisted, because *I knew* the information was there, and I just had to figure out how to extract it!

There is no unique recipe for what to do if we find anomalies, but we can develop some good practices.

Table 2.1: **The learning set for the habitable planets problem includes three features: the mass of the parent star, expressed in units of the mass of the Sun; the orbital period of the planet, in days; and the distance between the planet and its parent star, in astronomical units.**

| Name | Stellar Mass ($M_\odot$) | Orbital Period (days) | Distance (AU) | Habitable? |
|---|---|---|---|---|
| Kepler-736 b | 0.86 | 3.60 | 0.0437 | 0 |
| Kepler-636 b | 0.85 | 16.08 | 0.1180 | 0 |
| Kepler-887 c | 1.19 | 7.64 | 0.0804 | 0 |
| Kepler-442 b | 0.61 | 112.30 | 0.4093 | 1 |
| Kepler-772 b | 0.98 | 12.99 | 0.1074 | 0 |
| Teegarden's Star b | 0.09 | 4.91 | 0.0252 | 1 |
| K2-116 b | 0.69 | 4.66 | 0.0481 | 0 |
| GJ 1061 c | 0.12 | 6.69 | 0.035 | 1 |
| HD 68402 b | 1.12 | 1103 | 2.1810 | 0 |
| Kepler-1544 b | 0.81 | 168.81 | 0.5571 | 1 |
| Kepler-296 e | 0.5 | 34.14 | 0.1782 | 1 |
| Kepler-705 b | 0.53 | 56.06 | 0.2319 | 1 |
| Kepler-445 c | 0.18 | 4.87 | 0.0317 | 0 |
| HD 104067 b | 0.62 | 55.81 | 0.26 | 0 |
| GJ 4276 b | 0.41 | 13.35 | 0.0876 | 0 |
| Kepler-296 f | 0.5 | 63.34 | 0.2689 | 1 |
| Kepler-63 b | 0.98 | 9.43 | 0.0881 | 0 |
| GJ 3293 d | 0.42 | 48.13 | 0.1953 | 1 |

The above data set contains eighteen instances and three features: the mass of the parent star, expressed in units of the mass of our Sun; the orbital period of the planet, in days; and the distance between the planet and the star, in astronomical units (AU). One AU corresponds to the average distance between the Earth and the Sun, approximately 150 million km. The "Name" column is just meant to identify the examples and is not correlated with the target property, which is contained in the column "Habitable?". Note that it is tempting to just count the columns of a data set and assume this is the correct number of features, but alas, it is often wrong, as it would be in this example.

There are no missing data, and all the values seem to be fairly well behaved; perhaps the one that stands out is the orbital period of HD 68402 b, which, at 1103 days, is about one order of magnitude larger than any of the other values. A quick look at the other features suggests that this planet is also significantly farther away from its parent star compared to other instances of the data set, so it makes sense (remembering Kepler's laws) that it takes longer for it to complete one orbit. We can conclude that 1. This value is unlikely to be wrong; and 2. This instance might be statistically different from others (i.e., an outlier).

The data set is reasonably balanced, with 10 examples from one category (not habitable) and 8 from the other (habitable). Finally, we can discuss whether we expect these features to be informative enough to provide a reliable classification. At zeroth order, we can expect the answer to be yes; the single factor that determines whether a planet is habitable is its temperature, which will most likely depend on how much energy it receives from its parent star. This quantity is determined by a combination of the star's luminosity and its distance from the planet, so if we expect the mass of the star to be a decent tracer of its luminosity (as is the case for main sequence stars), we should expect our features to have a high degree of correlation with the desired answer. However, we know even from our Solar System that the reality is more complicated: The energy budget of each planet also sensitively depends on other features, such as the properties of its atmosphere, and whether it has an internal energy source. For example, on the basis of our simple argument, Venus and Earth should have somewhat similar temperatures, while in reality, the atmosphere of Venus created a runaway greenhouse effect that caused its temperature to be much higher. Additionally, the mass/luminosity relationship is monotonic only for main sequence stars, which make up only about 90% of the total. Therefore, going into this problem, I would imagine being only cautiously optimistic about our results.

For our initial investigation, we can split our data set into training and test sets; we will begin by using the first 13 examples as a training set, and the last 5 as a test set. We will explore other possibilities in the following section. Note that normally, the train/test split choice happens at random, and here we are making a different choice merely for reproducibility purposes.

## 2.2 DECISION TREES

It's now time for us to build our first, very simple, classifier, and we will start from one of the most intuitive ones: the decision tree. Decision trees are remarkably simple and work by asking a series of questions. Similarly to what we do when we are playing a guessing game, we want to get to the answer using the smallest possible number of questions. Of course, some questions are more informative than others.

For example, imagine that we are playing "Guess Who," a game in which you need to correctly guess the character on your opponent's card, from a range of possible choices (see Figure 2.2). You would probably agree that a good first question would be something that cuts the board approximately in half, and in general, questions that apply to larger groups are better; for example, "Do they wear a hat?" is usually a better question than "Do they wear a green hat?" since the second question would only give us useful information about one example. However, even after

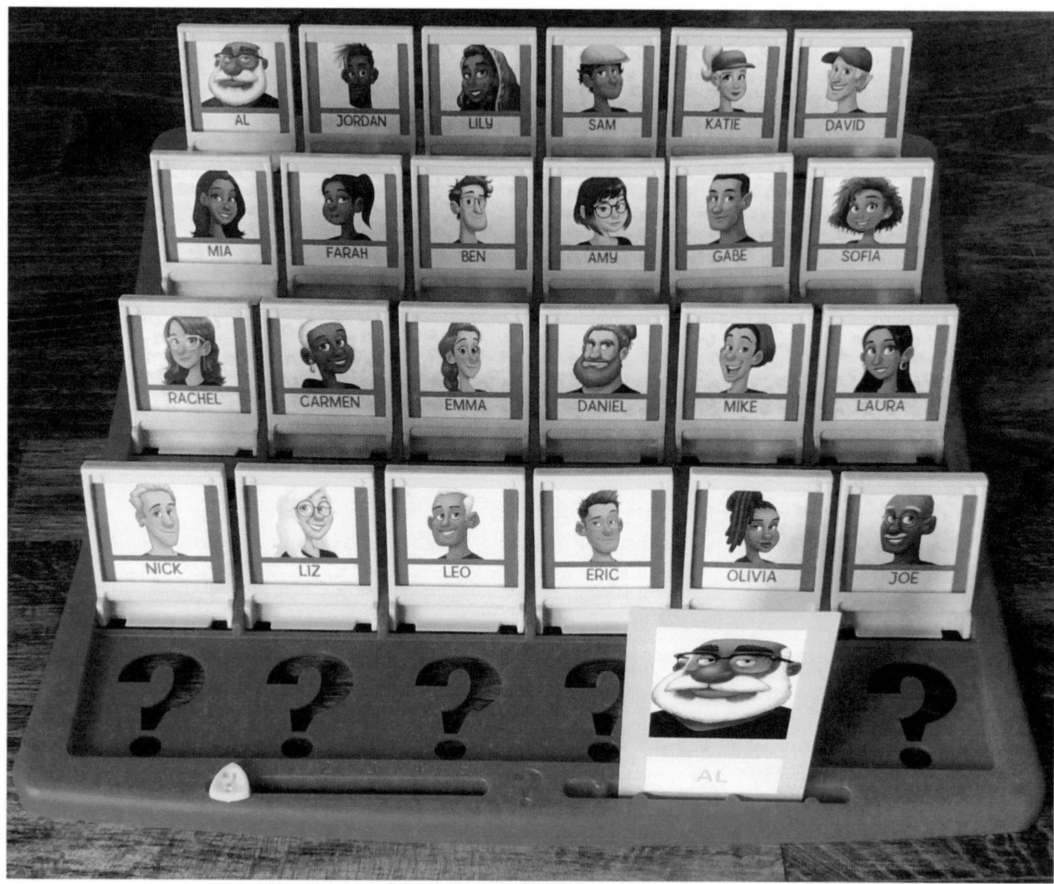


**Figure 2.2: The board of the popular "Guess Who" game.**

knowing whether our target wears a hat, we still won't know the answer, while in the odd chance that our target wears a green hat, we will know for sure that it's Katie! How to reconcile the desire to be accurate with increasing our odds of gaining useful information? We will see how we can summarize the *information gain* of each split by using the idea of *decrease of impurity* discussed in the next section.

### 2.2.1 How to build a decision tree

Let us now consider a simple example of how to use decision trees to solve a binary classification problem.

Decision trees are defined by splits (the criteria, or questions) and nodes (the groups of objects created by those criteria). Splits are binary, which means that the corresponding question can be formulated as a yes/no answer, or a true/false, or if it refers to a numerical feature, the question would take the form "Is the value smaller/larger than X?"

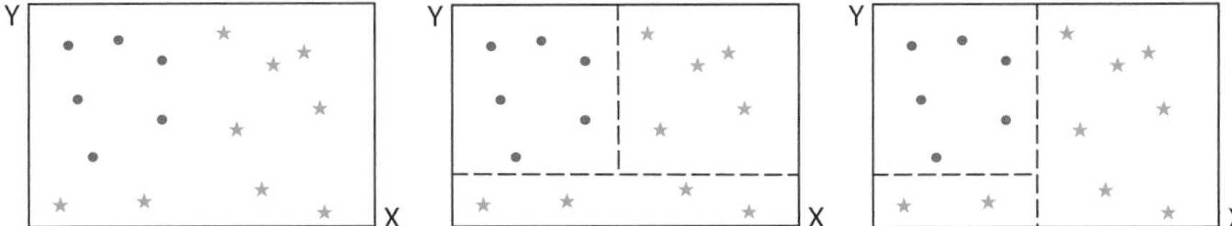

Figure 2.3: **A simple classification problem that can be solved equally well by two splits. But which one should we do first?**

Once we have implemented a split, we obtain two partitions, or nodes. If we are lucky, the two nodes are completely pure, meaning that they contain only objects of one type. But often that won't be the case, so we will implement further splits, obtaining more *leaf nodes*, until they are all pure (or until we decide to stop for other reasons, as we will see in Chapter 3). The terminal nodes of the tree are called leaf nodes, or simply, leaves.

### 2.2.2 Picking the splits

A good decision tree is characterized by efficient splits, so we need a mathematical criterion to decide which split is the best. For example, in the data set shown in Figure 2.3, it is clear that we can solve the classification problem using two total splits, roughly corresponding to $X > 0.5$ and $Y > 0.2$ (if we assume that the feature range is 0–1 for both X and Y). But which split should we do first?

We can formalize the concept of information gain, or maximum decrease of impurity, by using various criteria. One of the most common ones for classification problems is the so-called *Gini impurity*. The Gini impurity is defined as

$$1-\sum_i f(i)^2, \tag{2.1}$$

where $f(i)$ is the fractional abundance of each class. In a two-class problem, the Gini impurity varies between 0 and 0.5, with 0.5 being the highest level of impurity, corresponding to a set with equal numbers of objects from each class.

The Gini impurity of a proposed split can be calculated as the sum of the impurities of the two resulting nodes, weighed by the fractional volume of each node with respect to its parent node.

We can understand this better by looking at the example above (Figure 2.3). The original data set contains 6 dots and 9 stars, for a total Gini impurity of

$$1-\sum_i f(i)^2 = 1-(6/15)^2-(9/15)^2 = 0.48. \tag{2.2}$$

If we implement the horizontal split first (middle panel of Figure 2.3), we obtain two nodes, one with 4 stars and a Gini impurity of 0, and one with 6 dots and 5 stars, whose Gini impurity is $1 - \sum_i f(i)^2 = 1 - (6/11)^2 - (5/11)^2 = 0.496$. Therefore the total decrease of impurity achieved by this split is

$$\Delta\text{Gini} = 0.48\,(\text{original}) - 4/15 \times 0 - 11/15 \times 0.496 = 0.116. \tag{2.3}$$

Conversely, if we implement the vertical split first (right panel of Figure 2.3), we obtain two nodes, one with 7 stars and a Gini impurity of 0, and one with 2 stars and 6 dots, whose Gini impurity is $1 - \sum_i f(i)^2 = 1 - (2/8)^2 - (6/8)^2 = 0.375$. In this case, the total decrease of impurity is

$$\Delta\text{Gini} = 0.48\,(\text{original}) - 7/15 \times 0 - 8/15 \times 0.375 = 0.28. \tag{2.4}$$

The second option is preferable, because it achieves a larger decrease of impurity, as you have probably already figured out.

### 2.2.3 Predicting planet habitability with decision trees

We can now attempt to build a decision tree to predict planet habitability from the three features of the data set described in Section 2.1.2. We will use the first 13 examples as a training set, and the last 5 examples as a test set. I strongly recommend doing this with pen and paper first, by looking at the training set only and trying to write down the ideal splits. Then you can apply your tentative tree to the objects in the test set, figure out how many classifications you got right, estimate the test error of your algorithm, and compare it to the solution proposed by using the Decision Tree method in `sklearn` [Pedregosa et al., 2011].

Assuming that you have already done this, we can go ahead and follow the guidelines from Section 2.1. The lecture notebook "Intro_DT_HabPlanets.ipynb" provides an example on how to read the learning set of Table 2.1 from the file into a structure called *data frame* using the `pandas` library [McKinney, 2010], and how to split it into a training and a test set, as discussed in Section 2.1.2:

```
LearningSet = pd.read_csv(`HPLearningSet.csv')
TrainSet = LearningSet.iloc[:13,:]
TestSet = LearningSet.iloc[13:,:]
```

In this case, the file with the learning set contains both features and labels, so we have to separate them out in order to create the four arrays that are typically fed to `sklearn` algorithms: Xtrain (or training features), ytrain (training labels), Xtest (test features), and ytest (test labels). Of course, the files can be named anything you want.

```
Xtrain = TrainSet.drop([`P_NAME',`P_HABITABLE'], axis = 1)
Xtest = TestSet.drop([`P_NAME',`P_HABITABLE'], axis = 1)
ytrain = TrainSet.P_HABITABLE
ytest = TestSet.P_HABITABLE
```

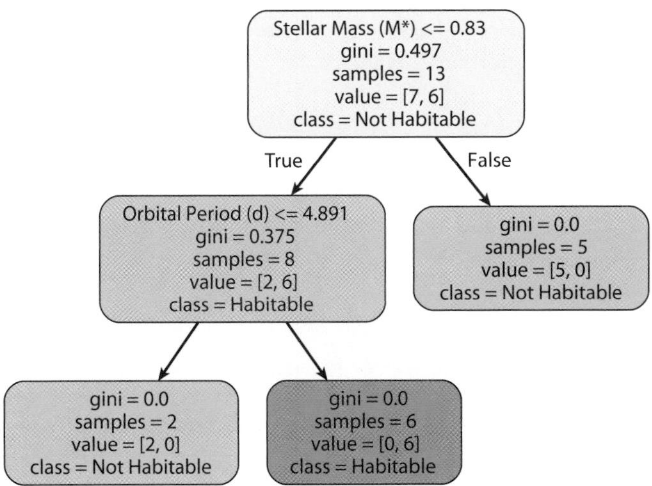


**Figure 2.4: The decision tree generated by the Decision Tree Classifier with default parameter values for the training set (first 13 rows) of Table 2.1. The intensity of the background color corresponds to the Gini impurity of each node; pure nodes have the strongest color. Each box shows sample size of the current node, the true sample distribution, the Gini impurity, and the current classification, which coincides with the final classification for leaf (terminal) nodes.**

We are now ready to import the Decision Tree Classifier model from `sklearn` and build the decision tree (keeping our random state fixed to make sure that results are deterministic), using the "`fit`" method:

```
model = DecisionTreeClassifier(random_state = 3)
model.fit(Xtrain, ytrain)
```

The decision tree model built by running the code above is illustrated in Figure 2.4.

A look at the figure tells us immediately that using just two of the three features, the mass of the parent star and the orbital period, and making two splits, we can generate a tree that outputs perfect predictions on the training set. Each leaf node is 100% pure, which means that the classifier's decision is aligned with the true label of every object in the training set.

However, when we apply the decision tree that we have obtained to the five objects in the test set, we see that two examples (HD 104067 b and GJ 4276 b) are misclassified as habitable, when in reality they aren't (see Figure 2.5). Therefore, we would conclude that even though our algorithm can classify 100% of the training set examples correctly, it only achieves 60% correctness on the test set. The percentage of correct classifications is often used as an evaluation metric of classification algorithms and is known as *accuracy*.

## 2.2.4 The impact of training/test set selection

Before we feel too sad about the less-than-perfect generalization of our algorithm, it is worth checking whether our chosen train/test set split had an effect on the performance of the algorithm. We can repeat our exercise of building the tree using the last 13 examples as a train set, and the first 5 as a test set (we could—in fact,

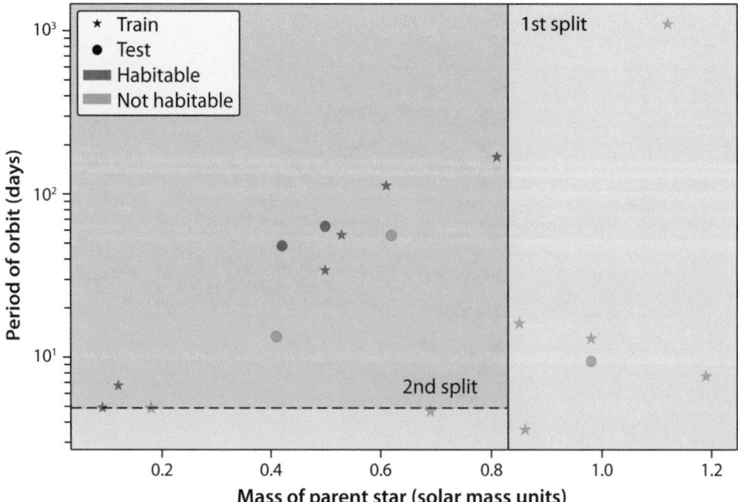


**Figure 2.5: The classification criteria from the tree of Figure 2.4 generate a perfect classification on the training set (stars) but fail for two test set examples (circles). The background colors indicate the prediction of the model, so a misclassification is indicated by a mismatch between a marker's color and the background color.**

should—also experiment with the size of the train/test set, but that will be left for the exercise section, and we will discuss it in Chapter 3).

Following the same approach of reading/fitting and predicting as above (see notebook "Intro_DT_HabPlanets.ipynb"), we obtain a new decision tree.

There are some obvious differences between this decision tree and the one generated previously. First of all, while it is still true that all the leaf nodes are 100% pure, as we can see from the bright colors of Figure 2.6, the new tree is deeper: It takes up to four splits to achieve this result. In other words, the decision-making process is more complicated. Second, the actual criteria used for the splits are different. Finally, in this case, the tree is using all three features to make a final decision. Based on all these differences, it is easy to imagine that the same object might end up with a different classification in the two cases. We could confirm this intuition by plotting the decision splits just as we did in Figure 2.5. Because the new tree involves splits in all three dimensions, the plot would be in three dimensions, so we will use our coding tools instead.

Which tree is better? To answer this question, we start by computing the accuracy on the test set, using the "`predict`" property of fitted `sklearn` models; you could also do this by hand by passing each object through the tree in Figure 2.6:

```
ypred = model.predict(Xtest)
print(metrics.accuracy_score(ytest, ypred))
```

The final result is that this tree achieves 100% accuracy on the test set. Does this mean that the second one is the "perfect" tree, or even just a better tree? We have to be cautious about making this judgment, because we are dealing with a very

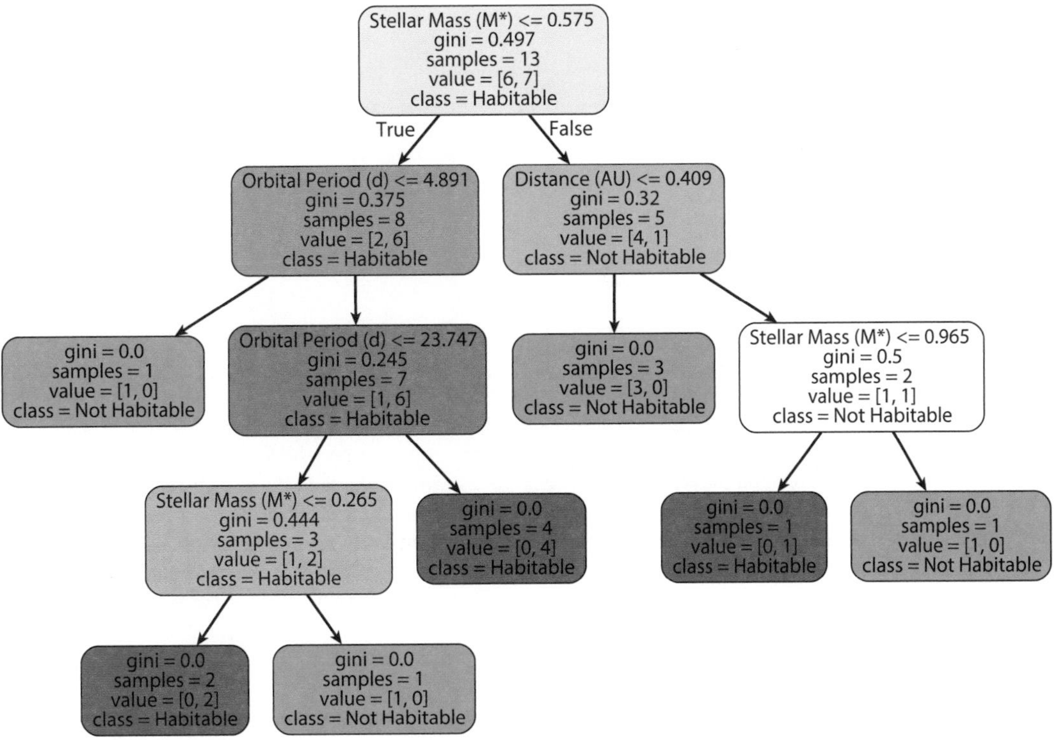


**Figure 2.6: The decision tree generated by the Decision Tree Classifier with default parameter values for a different choice of training set (last 13 rows) of Table 2.1. This train/test split generates a different, deeper tree than the one shown in Figure 2.4.**

small data set. The only sound lesson that we can learn for now from this exercise is that *different train/test splits might yield significantly different models, and expected test errors,* and thus this uncertainty should be taken into account when we assess the performance of our models. We will discuss this issue more in Chapter 3.

## 2.3 kNN: FINDING NEIGHBORS

Let us now consider the same planet habitability classification problem but use a different ML algorithm to solve it. Our goal here, besides familiarizing ourselves with different existing techniques, is to build an understanding of the strengths and weaknesses of each method or family of methods. This is useful for choosing methods that might be promising or learning how to watch out for potential pitfalls in our setup.

The $k$ Nearest Neighbors (kNN) algorithm [Cover and Hart, 1967] is a simple method that can be employed in classification or regression problems, just like the decision trees earlier in the chapter. When asked to classify an instance, the algorithm will look for the $k$ (with $k$ being a predetermined integer) closest

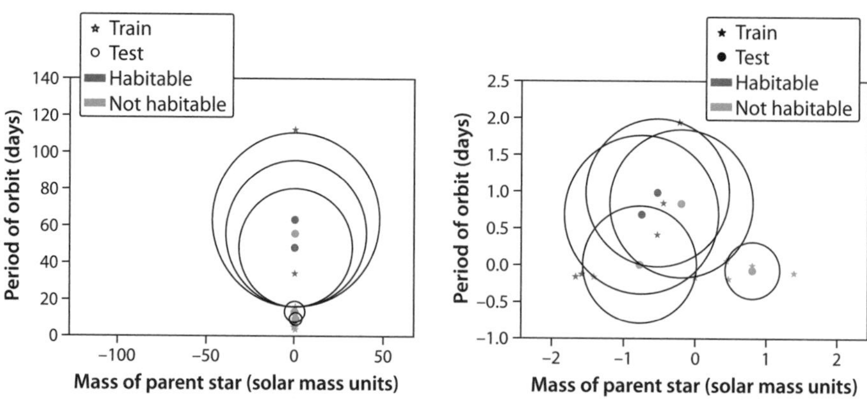


**Figure 2.7: The kNN classification algorithm finds the 3 closest neighbors from the training set (stars) around each test point (filled circles); they can be visualized as those that are inside each circle. On the left, we see what happens without rescaling the data; the distance measurement is dominated by the feature with largest numerical values, in this case the orbital period (note that the outlier in the training set is left out of the plot for better readability). On the right side, we see the effect of employing a scaling algorithm robust to outliers to bring both features to a similar numerical range. In this case, the classification of the test set points doesn't change, but this invariance is not common.**

examples in feature space in the training set and report as a decision the predominant class among those $k$ examples. In a binary classification problem, ensuring that $k$ is an odd number is a natural way to handle possible ties.

We can apply the kNN algorithm straight away to the Planet Habitability Lab data and compare its performance to the decision trees that we have just built. Because our training set is very small, it makes sense to use a smaller number of neighbors than the standard choice in `sklearn` ($k = 5$); let us set $k = 3$ for the moment and revisit this choice when dealing with parameter choice and optimization later.

We begin by applying the algorithm to the learning set of Table 2.1, using the first 13 examples as a train set and the last 5 as a test set, just as we did before. The practical implementation of the kNN algorithm in `sklearn` can be found in the notebook "Intro_kNN_HabPlanets.ipynb," but the usage pipeline is very similar to what we used for decision trees, with a call to the "`fit`" method on the training set to build the model and one to the "`predict`" method on the test set. To be able to visualize the results easily, we will use only the first two features, the stellar mass and the orbital period, but the results are very similar when all three features are used.

The results of the kNN algorithm are shown in the left panel of Figure 2.7. The black circles encompass the three closest neighbors in this two-dimensional space for each test set object (small blue and magenta circles). With a bit of squinting, you can probably figure out that the only incorrect classification on the test set is HD 104067 b ($M_* = 0.62\ M_\odot$ and orbital period $= 55.8$ days), for which two of the three closest neighbors are habitable planets. An accuracy of 80% on the test set is not bad, but we already know that due to the small size of the data set, we should not take these numbers too seriously. Instead, we can focus on another aspect revealed by the figure. The two features that we used for the classification, the stellar mass and the orbital period, have significantly different ranges, with a median value of 0.61 and 14.7, respectively. This might not seem like a big deal, and in fact, it was not an issue for our decision trees, because the algorithm is splitting on one feature at a time, so the scales of each feature are independent of one another. However, a method like kNN, which relies on distances computed in feature space, will be sensitive to different scales: In the left panel of the figure, we can see that the differences in stellar mass are basically negligible, which means that the algorithm is making its decisions based on the feature with the largest numerical values. This unwanted behavior can be solved by renormalizing the features and bringing them to a similar numerical range, as we see in the next section.

### 2.3.1 The importance of being standard

The kNN is one of many algorithms that solve problems by calculating some form of distance between instances (as opposed to, for example, decision trees, in which features are only considered one at a time). For such algorithms, *rescaling* (often referred to as *standardizing* in the ML community) the data before plugging them in a mathematical model is important, so that all features are given similar weight (we could of course also use the same mathematical tools in order to establish a hierarchy of features instead).

In general, one desirable outcome is to have all feature values be in the same numerical range and be "somewhat" normally distributed (i.e., when plotted in a histogram, the values look like a Gaussian); typical choices are a mean (or median) value close to 0 and a variance close to 1. This process happens for each feature independently. A nice overview of the effect of different standardization techniques can be found here.[4] The one-sentence summary is that different methods will be suitable to standardize different distributions, and sometimes it is good to experiment with more than one standardization procedure.

The simplest version of standardization involves calculating the mean and standard deviation of each feature (i.e., in each column), and modifying all values in that column by subtracting the mean and dividing the result by the standard deviation. However, the mean of a distribution is easily skewed by a few very large or very small values, and therefore this scaling technique (like any technique that revolves around mean values) is quite sensitive to outliers, such as the object with the very large orbital period in our data set. We adopt a more robust scaling technique that sets the median (instead of the mean) to zero and the difference between the 25th and 75th percentile (instead of the standard deviation) to one. In practice, this is achieved by creating a scaler object from the different options provided in the `preprocessing` class in `sklearn`, fitting it to the training set to derive

4 https://scikit-learn.org/stable/auto_examples/preprocessing/plot_all_scaling.html

the desired scaling transformation, and then applying this transformation to the training set and test set:

```
scaler = sklearn.preprocessing.RobustScaler()
scaler.fit(Xtrain)
scaledXTrain = scaler.transform(Xtrain)
scaledXtest = scaler.transform(Xtest)
```

Note that any preprocessing step (including normalization and standardization) should be done *on the training data only*, since using the statistical properties (like mean or variance) of the entire data set would introduce leakage of information from the training to the test process. Another way to think about this is that the test set can never be "seen" at any step during the model building process. Using for reference the first 13 examples as a training set, we obtain the plot on the right hand side of Figure 2.7, where we can see that the distance between neighbors is no longer dominated by one of the dimensions.

The kNN algorithm can then be applied to the scaled version of the training and test features. Note that in this particular case, the final classification accuracy of the test set points doesn't change, but this is an anomaly that depends on the small size of this data set, as we will see in Chapter 3.

## 2.4 LESSONS LEARNED

Let us conclude this chapter by summarizing some of the lessons we learned and summarizing some open issues that we will attempt to solve in Chapter 3.

- The first step of any project should be examining the data. Preprocessing, such as standardizing features, removing outliers if necessary, and deciding how to handle sparse data, is essential for the success of our endeavor.
- Different choices of training and test set split might yield different test errors (and, therefore, estimates of the generalization error). We need to address this issue by obtaining a measure of the typical performance and providing an estimate of the associated uncertainty. If the fluctuations are very high, we might consider this an indication that the learning set is too small.
- Decision trees are simple to interpret, powerful, and immune to different numerical ranges in the features, because features are always considered one at a time. An arbitrarily deep decision tree will always achieve a perfect performance on the training set. However, very deep trees might have issues in generalizing, and decision trees can only make splits that are aligned with the original features, suggesting the need for feature engineering before the model is built.

- The $k$ Nearest Neighbors (kNN) algorithm is another simple, interpretable, and fast algorithm for classification or regression problems. It is sensitive to the scaling of the feature, and it requires careful standardization if all features are supposed to participate equally. The algorithm will be sensitive to the size and density of the training set; if only a few training examples are available, or if only a few lie in the vicinity of a test set object, the algorithm would make decisions based on faraway examples. Some of these issues might be alleviated by considering neighbors that are within a fixed radius, as opposed to a fixed number of neighbors.
- Decision trees are able to prioritize different features: If some columns in our data set don't bring any information, they will simply be ignored, because splits along those directions won't decrease Gini impurity. However, kNN models will assign the same importance to all features, because distance is calculated over the full dimensionality of the data. Besides the scaling issue discussed above, we should keep in mind that redundant or useless features may harm kNN models, making them less accurate, because of the lack of a "feature prioritization" mechanism.

In addition to these considerations, thus far, we have only spoken of one evaluation metric, the accuracy, which is the percentage of correct classifications. There are other possible choices that might be more suitable for different problems; we will look at them in Chapter 3, which will be mostly devoted to evaluating and diagnosing supervised classification models.

Additionally, any ML algorithm will have some tweakable parameters that can be adjusted; for example, the maximum number of splits allowed in a decision tree or the number of neighbors to use in the kNN method. The issue of parameter tuning and optimization will be addressed in Chapter 4.

## 2.5 REVIEW AND DISCUSSION QUESTIONS

Note: Questions and exercises marked by ** are more complex, open-ended, or time consuming.

**Exercise 1.** I often play (actually, used to play, given how long it has taken me to finish this book) with my daughter "Guess the Princess." Please don't judge! Based on Figure 2.8, what is the best first question to ask (in terms of highest information gain)?

**Exercise 2.** What would be the feature values of the Earth-Sun system, in the reference frame of this chapter's problem? Would it be similar to the other objects or an anomaly?

**Exercise 3.** ** For the same problem and data set, if you were to engineer a different feature (e.g., a polynomial combination of existing features), which one would you pick and why?

**Exercise 4.** Do you think that feature engineering as proposed in the previous two exercises would have more impact on the Decision Tree algorithm or on the kNN algorithm?

Figure 2.8: Some princesses to guess.

**Exercise 5.** Would you recommend increasing the number of neighbors used in the classification for this chapter's data set? Why or why not?

**Exercise 6.** Another option for the kNN algorithm is to use neighbors that are within a certain radius, as opposed to a fixed number of neighbors. Would this choice help in this example? Why or why not?

**Exercise 7.** Using the figures in the text as a visual aid, what would happen to the test set error for different choices of the size of the train/test set split (e.g., 16/2 or 9/9)?

**Exercise 8.** Figure out the maximum Gini impurity in an N-class classification problem (you can assume that the maximum impurity occurs when objects are equally distributed among classes).

**Exercise 9.** If you used the two decision trees of Figures 2.4 and 2.6 to classify the Earth as habitable or not habitable, what verdict would you obtain? What can you conclude from this result?

**Exercise 10.** What is the purpose of splitting a learning set into train/test sets?

A) To deal with smaller data sets, which are more manageable
B) To avoid being overly optimistic by evaluating models on samples that have been used in the training phase
C) To improve the test scores
D) To reduce the number of features used

**Exercise 11.** What algorithm would you pick among the following options?

A) The one with the highest training scores
B) The one with the highest test scores
C) The one with the highest difference between test and training score
D) The one with the lowest difference between test and training score

**Exercise 12.** Which of the following is the best proxy for the generalization error of a model?

A) The training score
B) The test score
C) The training error
D) The test error

**Exercise 13.** What is the Gini impurity of a data set with two classes, where 2 objects belong to one class, and 6 objects to the other class?

A) 0
B) 0.375
C) 0.5
D) 0.625

**Exercise 14.** What is the Gini impurity of a data set with three classes, where 2 objects belong to the first class, 2 objects belong to the second class, and 6 objects belong to the third class?

A) 0.2
B) 0.4
C) 0.56
D) 0.8

## 2.6 PROGRAMMING EXERCISES

**Exercise 1.** Write the pseudocode for the Decision Tree algorithm (a sequence of operation, or control flow, in plain English), assuming that you use the Gini impurity as a decision criterion. This is a possible incipit to get you started:

```
Calculate Gini impurity of learning
set
IF stopping criterion is met THEN
generate prediction
ELSE ....
```

**Exercise 2.** Code up kNN from scratch, using the Euclidean distance and uniform weights.

**Exercise 3.** Redo the kNN exercise using the last 13 examples as the training set, and the first 5 as a train set. What is the accuracy on the test set?

**Exercise 4.** Check how the performance would change if you chose different sizes for the train and test set.

**Exercise 5.** Redo the kNN exercise using the "`StandardScaler`" option instead of the "`RobustScaler`" one. What happens to the test set accuracy and why? (Suggestion: Plot each point and their neighbors.)

# CHAPTER THREE

# 3 Supervised Classification: Evaluation and Diagnostics

*What's Marie Kondo's favorite cross validation technique?*
*k-fold.*

## 3.1 WORKING WITH RESEARCH-LEVEL DATA SETS: PREPROCESSING AND ANALYSIS

It is now time to step up our game and consider a data set that is more similar to what everyday research data sets look like: larger, messier, and noisier than what we saw in Chapter 2. In fact, we will still work with the Planet Habitability Lab data set, but we will now consider the entire sample (and I can now confess that the small selection used was carefully selected to be "first-step friendly"). This version of the data set was downloaded at writing time, in January 2020, so it is possible that if you head to the website, you will now find even more exciting exoplanets. At this time, the data set contains data for 4,048 planets, and it hosts information well beyond the three features we have been using; there are in fact a total of 97 numerical features and more than a dozen categorical features. However, our purpose is to predict habitability from features that are easily measurable, and therefore we will keep using only the stellar mass of the parent star, the distance between star and planet, and the orbital period. Furthermore, a closer look at the data set reveals that the planet habitability label was not binary, but had three possible values: 0 (for not habitable), 1 (for optimistically habitable), and 2 (for reasonably expected to be habitable). For our exercise, we will combine the "1" and "2" as 1 (habitable), but we will discuss nonbinary classification at the end of this chapter.

We begin our exploratory data analysis looking for missing data and other statistical oddities (see the lecture notebook "HP_Chap3.ipynb"). In this larger data set, many measurements are missing. In particular, there is no stellar mass value for 765 objects out of the 4,048 total, while the orbital period and distance measurements are unavailable for 110 and 70 objects, respectively. How should we deal with this problem?

There is no unique strategy to handle missing values. The most draconian option is to discard objects for which any of the features are missing, but this of course implies throwing away data, which is never ideal. Another possible choice is to replace the missing value with some plausible estimate (called *imputing* in ML jargon), for example, the mean or median of that feature, or some plausible value coming from building a predictive model of the missing values (e.g., with a kNN!), using the other features as input (see Section 4.3.1). Of course, the larger the fraction of data we are replacing with estimates, the larger the overall effect will be, and the more careful we need to be. We will adopt the first technique here:

```
final_features = final_features.dropna(axis = 0).
```

This approach eliminates all instances that have at least one "not a number" (NaN) in any column; I invite you to consider other possibilities in the exercise section. Using this approach curtails our data set to 3,180 objects.

A first numerical analysis of the features reveals that the distributions are quite skewed, and there are several outliers. These characteristics can be shown by plotting each feature as a histogram or just by using the "`describe`" property of data frames in `pandas` and noticing that the mean and median of each distribution are quite far from each other. In this case, let us opt to eliminate severe outliers, which might "skew" decisions made by our classifiers, by excluding objects whose z-score (distance from mean, calculated in units of standard deviation) is larger than 5. There are of course many other possible choices, and which one is most appropriate depends on the problem, as well as on the algorithm, because different algorithms have different responses to outliers.

We are now ready to build our models on a revised data set of 3,171 objects and three features. We plot the first two features, together with the true label, in Figure 3.1, which shows two main differences compared to the small version seen in Chapter 2. One, all features span a much larger range than before (one-to-two orders of magnitude). Two, the larger data set is much more imbalanced, with many more examples of nonhabitable planets than habitable ones (something that should probably not come as a surprise, given the example of our own Solar System). A quick count of elements in each class reveals only 52 habitable planets.

Overall, we can conclude that the statistical properties of the larger data set are significantly different from the ones of the small excerpt we have been using; as a result, trying to recycle one of the models we had previously built will presumably lead to a very poor performance (test it for yourself in exercise 2 in the Programming Exercises at the end of the chapter). This simple example illustrates two of the trickiest issues in generalizing ML algorithms: how to determine whether we have enough data, and whether we can trust models once we move away from the original domain where the models were derived. We will continue this conversation and develop our first diagnostic tools in Section 3.5.

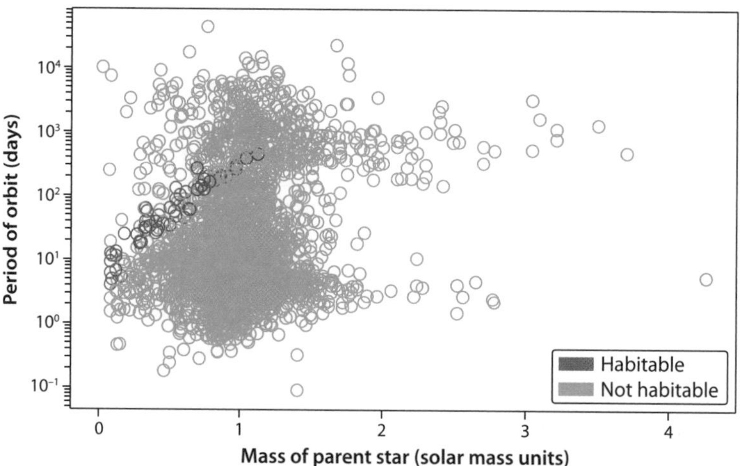


**Figure 3.1: The full habitable planet data set, as retrieved from the Planet Habitability Lab website of Arecibo Observatory in January 2020. Compared to the small data set used for the exercises in Chapter 2, this data set contains $\sim$ 175 times more instances, spans a much larger feature range, and is extremely imbalanced.**

## 3.2 BINARY CLASSIFICATION EVALUATION

So far, we have only considered accuracy, the total percentage of correct answers, as a possible evaluation metric for our models. But looking only at one possible metric does not provide a full picture of where our model is performing well and where it is failing, which might be deceiving.

For example, let's suppose that we build a very lazy model to predict planet habitability, which puts everything in the negative (not habitable) class. We can probably agree that such a classifier is rather useless, as it has not learned anything about what characteristics make a planet habitable or not habitable. What would be the accuracy of such a classifier? It would be the number of correct classifications $(3171 - 52 = 3119)$, divided by the total number of instances $(3171)$; in other words, a whopping 98.4%! This trait is known as the *accuracy paradox*: In an imbalanced data set, a very high accuracy is not necessarily indicative of an informative model.

We can build further understanding of model performance by noting that for a binary classifier, once we have assigned a "positive" label to one class and a "negative" label to another, any metric will be defined in terms of four numbers: the amount (or fraction) of true positives (TP), true negatives (TN), false positives (FP), and false negatives (FN).

In this convention, the second word (positive/negative) is *the classification assigned by the model*, and the first word indicates whether such classification is correct (true), or incorrect (false). I find it a bit confusing that the intrinsic label (ground truth) is not actually spelled out, so I tend to paraphrase these in my head as "correctly classified as positive," "correctly classified as negative," "incorrectly classified as positive," and "incorrectly classified as negative." Incidentally, the positive/negative label assignment is completely arbitrary, although typically the rare/interesting class is identified as positive, and it should be noted that not all evaluation metrics are symmetric with respect to the definition of positive/negative class.

In this language, accuracy is defined as the total number of correct classifications (the one with "true" as prefix) over the total number: (TN+TP)/(TN+TP+FN+FP).

We can now proceed to define some other common evaluation metrics: *precision* and *recall*.

Precision is the fraction of correct positive classifications. The total number of positive classifications is (TP + FP), so precision is defined as TP/(TP+FP). It answers the question: Out of all the examples that have been classified as belonging to the positive class, how many of those truly belong to the positive class? Recall is the fraction of correctly identified positive examples. The (intrinsically) positive examples can be obtained as the sum of the true positives, TP, and the false negatives, FN; therefore, the fraction of correct classifications is TP/(TP + FN). This fraction answers the question: Out of the totality of objects that truly belong to the positive class, how many of them has our model captured?

I find it useful to think of these two quantities, in plain English, as purity and completeness respectively. Note that unlike accuracy, precision and recall are not symmetric with respect to the choice of positive class (they would "flip" if we swap $F \rightarrow T$). For the same reason, when we calculate them using the built-in attributes in `sklearn`'s "`metrics`" package, we need to be careful about placing the ground truth first and the predictions second. This message is brought to you by many hours of frustrated debugging.

In the above example, the "lazy" classifier would have 0 true positives, 0 false positives, 3,119 true negatives, and 52 false negatives. We can easily calculate the precision and recall, which will be undefined (0/0) and 0 (0/52), respectively. By calculating a set of evaluation metrics rather than just one, we would be able to tell that something is amiss in our model.

A useful representation of the performance of a binary classifier is the *confusion matrix*, shown in Figure 3.2.

This matrix, which is $2 \times 2$ for a binary classifier, is built by comparing the predicted labels (typically plotted as columns) and the ground truth (typically plotted as rows). With these plotting conventions, correct classifications appear on the negative-slope diagonal of the matrix, while incorrect ones appear on the positive-slope diagonal; false negatives are reported in the lower left quadrant, while false positives appear in the upper right quadrant. The confusion matrix offers an intuitive visual way of capturing the performance of a classifier, and any metric can be easily recast in terms of its elements: Accuracy is the sum of diagonal terms over the

**Figure 3.2: The confusion matrix for the "lazy" classifier, indicating the location of true positives, true negatives, false positives, and false negatives. Predicted labels are shown in the columns, and ground truth labels are shown in the rows. With this convention, correct predictions lie on the negative-slope diagonal, while incorrect ones are on the positive-slope diagonal.**

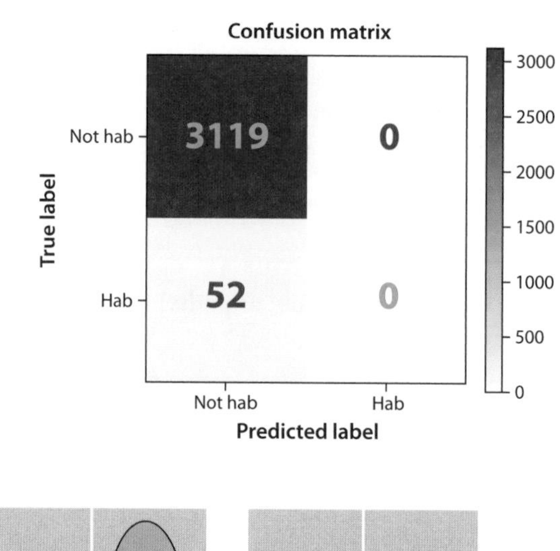


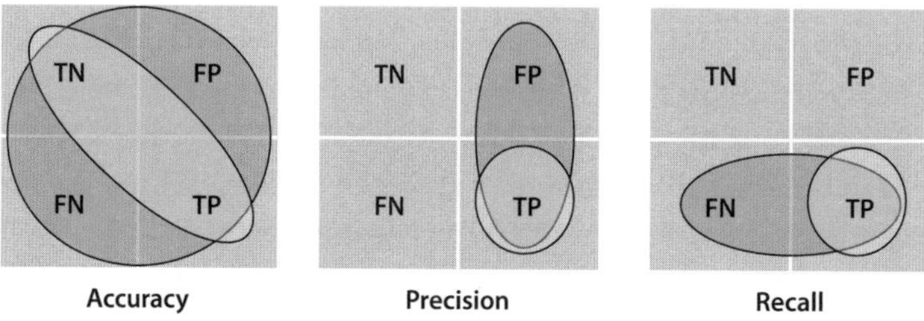


**Figure 3.3: Visual mnemonics for accuracy, precision, and recall can be obtained by dividing the numbers in the green clouds by the numbers in the blue clouds.**

total; precision is the lower right quadrant (true positives) divided by the sum of the terms in the second column (true + false positives); and recall is the lower right quadrant (true positives) divided by the sum of the terms in the second row (true positives + false negatives). These visual shortcuts are illustrated in Figure 3.3.

Other metrics can also be created from different combinations of the four elements of the confusion matrix. For example, the *F1 score* is a weighted combination of precision and recall.

### 3.2.1 Receiver operating characteristic curve and area under the curve

Another commonly used metric to evaluate the performance of a classification model and to compare different models is the receiver operating characteristic curve, or ROC curve (see, e.g., [Fawcett, 2006]). To build the ROC curve, we need to go back to the idea of recasting classification problems as regression problems introduced in Section 1.4.2. Remember that rather than just outputting a class we can train models to output the probability that an object belongs to a certain class and then adopt a probability threshold (typically 0.5 in a standard binary classification problem) to make the class assignment. This approach is advantageous because this regression model retains useful information about the confidence of our classification.

We can leverage the idea of a threshold to explore how the decisions made by a classification model change *as the threshold changes*. Let's assume that we now assign each object to the positive class if the predicted probability of belonging to the positive class is $\geq 0.1$, rather than the "standard" choice of 0.5. Thus, objects are very easily assigned to the positive class, and so the rate of TPs will increase (and as a consequence, the rate of FNs will decrease), as we will miss fewer of them, but of course the rate of FPs will also increase, as we include in our count even objects that were assigned to the positive class with very low probability. In the limiting case of a threshold of 0, all objects are assigned to the positive class, which brings the rate of TP to 1. Similarly, if we tune the threshold to a high value, say for example, $P_{\text{pos}} = 0.9$, only objects that were assigned to the positive class with very high probability will be classified as positive. Thus the rate of FPs will decrease, as we include fewer objects in our "positive" count, at the expense of an increase in the rate of FNs, as we are much more likely to "miss" objects that are truly positive by setting this very high bar to include them. Overall, varying the threshold has the effect of adjusting the balance between the two types of errors: FNs and FPs, and this in turn corresponds to trading off precision for recall. A low value for the threshold will induce low precision (because precision is sensitive to FPs), and a high value for the threshold will induce low recall (because recall is sensitive to FNs). This behavior further confirms that looking at only one metric (accuracy, precision, or recall) may be misleading (because we can artificially inflate precision or recall simply by varying the threshold), but it also opens up the idea of looking at the performance of a classifier *as a function of a varying threshold*, which is the idea behind the ROC.

An ROC curve is built by plotting the *rate of false positives* (FPR), defined as FP/(FP + TN), versus the rate of true positives (TPR), which is another name for recall, TP/(TP + FN). For different values of the threshold, varying between 1 (no object is classified as positive, so FP = 0 and TP = 0) and 0 (all objects are classified as positive, so TN = 0), the values of the curve vary between (0,0) and (1,1), and the curve usually ends up looking something like the diagram in Figure 3.4.

We can now discuss how the ROC of a good model would look. No matter the model, the initial and final points of the curve are set. Because of the definition, the curve will always be monotonically increasing: The true positive rate can only increase as more and more objects are included in the "positive" count, but the same is true for the false positive rate. For highly predictive models, however, the curve will rise sharply and approach the "perfect classifier" case, which has a close-to-zero FP rate and close-to-one TP rate (point (0,1) in the ROC diagram). For nonpredictive models, the curve will stay closer to the diagonal line. For this reason, the area under the curve (AUC) is often used to summarize the performance of a model and to compare different models across a range of possible choices of threshold (i.e., across the range of possible precision/recall trade-offs).

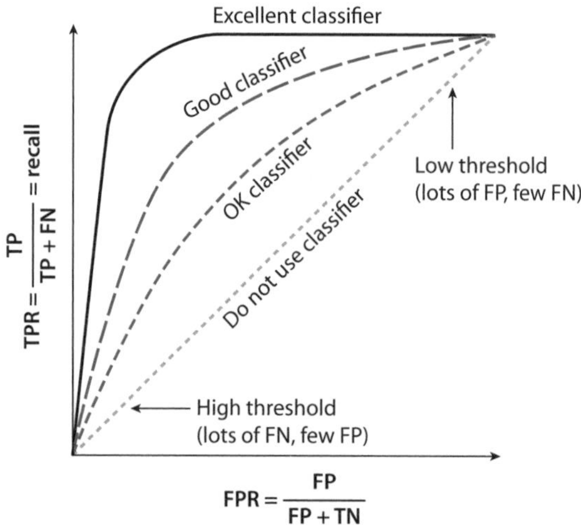


**Figure 3.4: Receiving operator characteristic (ROC) curve for different classifiers. The ROC curve plots the false positive rate (FPR) versus the true positive rate (TPR, or recall). For each model, different points along a curve correspond to different probability thresholds above which an object is assigned to the positive class. Note that the threshold decreases from left to right.**

## 3.3 CHOOSING AN EVALUATION METRIC

After getting acquainted with several possible evaluation metrics, we are better equipped for choosing the most suitable one for our problem. But which one is the right one? There is, of course, no universal answer. For example, in our case, accuracy is not a "bad" tracer in itself. However, unlike the other two metrics (precision and recall), accuracy is symmetric in the two error types (FNs and FPs), which means that the model will give equal weight to either class. In an imbalanced data set, this translates into a much higher *fractional* misclassification rate for the rare class. For example, 10 misclassified examples in the "nonhabitable" planet class are only $\sim 0.3\%$ of the total for their class, while 10 misclassified examples in the "habitable" planet class are 20% of the class total. We could then decide to choose one of the other metrics.

Maximizing precision is useful when we want high purity in the proposed sample of positive classifications. Precision is sensitive to false positives only, so a high precision ensures that each positive classification comes with high confidence. This happens at the cost of some positive examples being "missed" by the classifier or, in other words, a higher rate of FNs, which translates into lower recall.

Conversely, maximizing recall is often associated with the search for rare objects, in which we tend to care more about "catching" all possible positive examples by casting a wider net, which will minimize the number of false negatives and ensure completeness. The flip side is the resulting lower purity of our sample of proposed positive classifications (i.e., lower precision, or a higher rate of FPs).

For problems like the search for habitable planets, it is tempting to just use recall as our evaluation metric, because such planets are rare and we don't want to miss a place that potentially hosts intelligent life! However, in practice, the answer is more complicated, as it might depend on the total resources that we have to follow up (e.g., we may only have money to do direct imaging of the 20 planets that are most likely to host life), or on a combination of probability of hosting life with distance from us, and so on. Therefore, in most real-world scientific applications I have seen, it is common to define one's own metric, for example, a combination of the four numbers with different weights. We will return to the topic of custom loss functions in multiple examples in the remainder of this book.

## 3.4 BEYOND TRAINING AND TESTING: CROSS VALIDATION

Before we proceed to create a model for our larger data set, we need to improve our training/test splitting technique. We already know from Chapter 2 that different choices of training/test split might significantly affect the resulting performance, because of the variation in the statistical properties of the splits. We can expect this to be more significant for small data sets, whose statistical properties can be skewed by just a few elements, but it is relevant for large data sets as well. When we pick a single train/test split, we might significantly under-/overestimate the performance of our algorithm. Instead, we would like to quote a *typical* performance, quantified by our evaluation metric of choice, and to also provide an estimate of the associated uncertainty coming from the statistical variability of the data. Another shortcoming of the single train/test split approach is that we effectively waste a significant amount of potential training data, since the data in the test set are not allowed to be part of the model-building process. Given that data are often difficult and/or expensive to acquire, this aspect is also less than ideal. Both issues can be solved effectively by a process known as *cross validation*.

The general idea is that we choose several possible train/test splits and repeat the training process accordingly. The overall performance can be estimated as the mean (or median) of the test scores obtained in all the attempts, and the standard deviation (or other dispersion measure) provides an estimate of the uncertainty.

The most common practical strategy for cross validation is $k$-fold cross validation, which consists of dividing the learning set into $k$ folds, and "cycling through" the folds so that in each of the $k$ iterations, we use one fold as test set, and the remaining $k-1$ folds as the train set, as shown in Figure 3.5. This strategy is slightly preferable (although effectively equivalent, if the data set is large) to choosing $k$ random realizations of train/test splits, since it ensures that each sample in the data set is part of the training set exactly $k-1$ times, and of the test set 1 time, so that all samples have exactly equal weight. One technical but very important detail is that we need to make sure the data set is shuffled before employing this technique, because any ordering is preserved if the $k$ folds are chosen sequentially, which is the default option in `sklearn`.

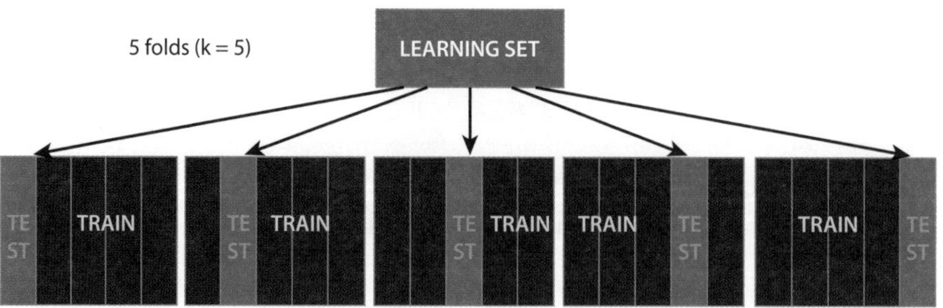


**Figure 3.5: In *k*-fold cross validation, we divide the data set into *k* parts or folds, and we use $k - 1$ folds as the train set, and one fold as the test set. We cycle through the test fold *k* times, so that each instance participates in the process equally. Compared to picking a single train/test split, this process provides a more accurate assessment of the test error, and an estimate of the associated uncertainty caused by the variance in the learning set.**

Through this process, we build a predictive model $k$ times. So which one is the right, "final" model? The answer, perhaps surprisingly, is *none of them*. The purpose of the cross validation process is to provide a fair estimate of the generalization error and its uncertainty. However, the final model will be built using the *entire learning set*, and therefore it might be slightly different from any of the models obtained previously.

## 3.4.1 How many folds?

The main disadvantage of cross validation is that we need to repeat the training process $k$ times, which can quickly become computationally expensive (especially once we introduce the parameter optimization process, as we will see in Chapter 4). Using a large number of realizations (folds) has multiple advantages: It makes the estimate of the mean/median and standard deviation more reliable, and it ensures that the generalization error for the final model, built on the full data set, is similar to the one reported in the cross validation (CV) process, in which the training set size is smaller. This matters especially for small data sets (see Section 3.5.3 for more details). However, the training time scales linearly with the number of folds, which might limit one's ability to use a large number for $k$. Therefore, the "right" $k$ for a given problem comes from establishing a trade-off between repeating the procedure many times and keeping computational times reasonable. Typical $k$ values are between 5 and 10; it is not advisable to go below 3 if at all possible.

## 3.4.2 Other CV techniques

Most other CV techniques are some variation of the basic $k$-fold method. One that has received a certain amount of attention is the *Leave-One-Out* CV, in which one considers as many folds as examples in the data sets, so that each training set has $n - 1$ objects, and each test set has only one object; a similar approach is taken by the *Leave-p-Out* CV method, where $p$ objects are left out for the test set, and the possible splits are permutations of $(n-p)$ objects. The User Guide of `sklearn` [Pedregosa et al., 2011] contains a useful comparison of different CV techniques and their relative advantages and disadvantages.

One further "flavor" of $k$-fold CV that is relevant for imbalanced classification problems is *stratification*, the process of choosing the folds while maintaining approximately the same fractional class membership of the entire learning set. For example, if the ratio of negative:positive class members is $1/20$ overall, a stratified $k$-fold partition will keep approximately the same ratio in all the folds. This tends to occur naturally as long as the folds are chosen randomly (or equivalently, the data set is shuffled before segmentation), but if the imbalance ratio is strong and the data set is small or the number of folds is large, we might end up with training/test folds whose statistical properties differ significantly from those of the entire data set. In these cases, it is beneficial to use stratified CV strategies.

### 3.4.3 Implementing cross validation

The `sklearn` library contains a series of convenient functions to implement different CV strategies and evaluate the corresponding results. In our case, our data set is severely imbalanced, and therefore we will use a stratified $k$-fold approach, comparing the results for the two algorithms we have introduced so far (i.e., decision trees and kNN). This model is very simple, so we can safely use 10 CV folds. We will fix the random state both in the CV strategy and the model to ensure reproducibility:

```
cv = StratifiedKFold(shuffle = True, n_splits = 10, random_state = 10)
```

(All the code examples in this chapter come from the lecture notebook "HP Chap3.ipynb.") Note that we made sure our data set is shuffled before applying the CV splits. The next step is to use the `cross_val_score` function (or equivalently, the `cross_validate` function), which splits the data according to the CV strategy specified above and reports the performance, evaluated via the "scoring" parameter, in a vector of length $k$:

```
scores = cross_val_score(model, final_features, targets,
cv = cv, scoring = 'accuracy')
```

For the Decision Tree Classifier model, the mean and standard deviation of the scores vector are 0.983 and 0.005, respectively: in other words, this classifier is not better than the "lazy" classifier that puts everything in the non-habitable category. We can further investigate the behavior of our model by looking at its precision and recall values as we adjust the scoring parameter above; for the same setting, we obtain a precision value of $0.53 \pm 0.2$ and a recall value of $0.5 \pm 0.13$. In other words, the model is only able to correctly identify about half of the positive class members, and out of the proposed positive classifications, only half of them are

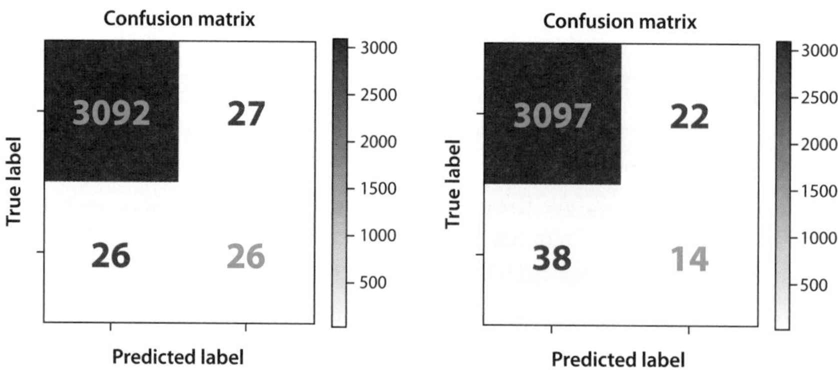


**Figure 3.6: Confusion matrix for the Decision Tree Classifier (left) and kNN classifier (right) with default parameters.**

correct. For a more comprehensive summary, we can take a look at the confusion matrix of this model. There is no "ready" way of averaging the confusion matrix in all the folds, as we did for the scores, so we can do it by hand or use the shortcut of the `cross_val_predict` function, which gives a visualization of the labels obtained by assembling the predictions on the 10 test folds (so it is one possible *realization* of confusion matrix but doesn't carry any notion of uncertainty):

```
pred = cross_val_predict(model, final_features, targets, cv = cv).
```

Comparing the vector of true and predicted labels gives us the confusion matrix in the left panel of Figure 3.6, with 3,092 TNs, 27 FPs, 26 FNs, and 26 TPs.

We can do the same exercise using our kNN classifier, remembering that we have to standardize our data before applying the algorithm. One important caveat already mentioned is that the standardization process needs to happen on the training set only, to avoid leakage of information from the test set to the model (e.g., [Hastie et al., 2001]). This applies also when using a CV strategy, so we can use a "pipeline" method to ensure that standardization is applied to each of the train/test splits used inside the CV object:

```
scaler = sklearn.preprocessing.RobustScaler()
model = KNeighborsClassifier()
pipeline = Pipeline([('transformer', scaler), ('estimator', model)])
```

We can then use the `cross_val_score` and `cross_val_predict` functions, using `pipeline` as the first argument (so that our model is, in fact, the combination of standardization and predictive modeling).

The kNN algorithm with default parameters performs worse than the decision tree, with recall and precision scores of $0.27 \pm 0.23$ and $0.32 \pm 0.22$, respectively. The confusion matrix stemming from the `cross_val_predict` procedure is shown in the right panel of Figure 3.6, and shows 3,097 TNs, 22 FPs, 38 FNs, and 14 TPs.

# 3.5 DIAGNOSING A SUPERVISED CLASSIFICATION MODEL

The confusion matrices shown above demonstrate that the two algorithms we have so far used are not performing ideally. The next step is *diagnosing* the algorithm(s) by gathering more insights into what is going wrong and understanding what steps can be taken to improve performance.

The very first diagnostic step, which we have in a way introduced already, is data exploration. Such factors as the presence of strong outliers, missing data, features with widely different magnitudes, and severe imbalance are often the root cause of less-than-satisfactory performance. By considering each factor carefully, we can understand what algorithms or sampling strategies might be most helpful for our problem.

Beyond this first step, other diagnostic tools can help reveal where there is room for improvement.

## 3.5.1 Overfitting and underfitting

ML models tend to suffer from one of these issues: *underfitting* or *overfitting*; often these problems are associated to the concept of high bias and high variance, respectively.

A model that suffers from high bias cannot capture the complexity of the input-output relationship (in other words, it's too simplistic). It might be because of the algorithm that we chose (e.g., we are using a linear model to capture a nonlinear relationship), or because the features that we have are not relevant or informative enough, or because we don't have enough data. Whatever the reason, the practical consequence is that our test scores are low.

A model that suffers from high variance generalizes poorly, because it has been excessively tailored to its training set. This often happens when the model is excessively complex; perhaps it's trying to fit simple input/output relationships with highly flexible methods, with the result of "absorbing" all the nooks and crannies of the training set, including random fluctuations and outliers (e.g., see the right panel of Figure 4.7). Another possibility is that it's been fed a lot of information (features) that are not helpful and act as noise. Whatever the reason, the practical consequence, just as above, is that our test scores are low.

Given that both problems might induce low test scores, how can we diagnose whether we suffer from one or the other? A useful diagnostic tool is to look at the difference between *train* and *test* scores. If the model suffers from high bias, train and test scores will be similar, because the inability of the model to capture the correct relationship between features and target applies equally to training and test examples. But if the model suffers from high variance, there will be a large gap between training and test scores, because the model mistakenly retains information that only pertains to the training set. We will return to this topic and show a more formal analysis of bias/variance decomposition in Section 5.5.

It is important to note that identifying a high variance or high bias issue is not a solution but a diagnostic step: Identification just offers insights into where there might be room to improve the model, so that can we try some of the strategies discussed next.

### 3.5.2 Diagnosing high bias and high variance in practice

We can take a look at our two algorithms, the decision trees and the kNN, and try to understand whether the low scores we observe (e.g., precision/recall values in the 50–60% range) are due to high bias or high variance.

To do so, we need to take a look at the performance of the algorithms on the train set and on the test set, in a cross-validate-y fashion. The simplest way to do so in `sklearn` is to use the `cross_validate` function, with the option of retaining the train scores. If we choose recall as our scoring parameter, we have

```
scores = cross_validate(pipeline, final_features, targets, cv = cv, scoring =
`recall', return_train_score = True).
```

Here the pipeline model setup ensures that scaling is implemented within the train fold only, and the CV strategy is still the stratified $k$-fold with $k = 10$.

For the DT model, the mean and standard deviation on the train set and test set are $1.0 \pm 0.0$ and $0.5 \pm 0.13$, respectively. We knew this already from the discussion earlier in this chapter: A decision tree with an arbitrary number of splits will keep splitting into smaller and smaller nodes until it reaches 100% accuracy/precision/recall on the train set. Therefore, we observe a large gap between train and test scores, and the estimate of the standard deviation tells us that the gap is highly statistically significant (if we believe that the distribution of scores is Gaussian, this would be a $4\sigma$ significance). We would need to improve the DT model by finding ways to reduce variance.

If we do the same exercise on the kNN algorithm, we find that the mean and standard deviation on the train set and test set are $0.5 \pm 0.04$ and $0.27 \pm 0.23$, respectively. This model suffers from high bias, because scores are low both on the train and the test sets. Despite the apparent large gap between train and test scores, the large uncertainty means that we can't conclude that it is statistically significant yet, and so we should take steps to correct the bias before we take steps to correct the variance.

A final thought on this subject is that a particular model can suffer from *both* high variance and high bias; our kNN algorithm is a borderline example of this behavior. In the following sections, we will look at some possible strategies to fix each problem. Some of them are based on the idea of bias-variance "trade-off"; in other words, we assume that the total generalization error of our models depends on a combination of bias and variance, and we aim to decrease it by improving one of the issues at the expense of the other. A longer discussion of the bias-variance decomposition and its interpretation in terms of model complexity is given in Chapter 5. In general, understanding the roots of our models' behavior will be helpful in deciding which strategies are most helpful for our case.

### 3.5.3 Learning curves

Learning curves are another useful diagnostic tool for supervised models. They are used to estimate how the performance of the model is tied to the size of the learning set. Intuitively, if our learning set is very small, models won't be able to capture the input/output (I/O) relationship correctly and will have low test scores. As we add more data, their performance would improve until a plateau is reached when the data have enough variance to maximize the information available from the provided features. After this point, adding more data (in the sense of adding more data with identical features and distribution of the current data) won't help any more, indicating that we need to look elsewhere to improve our scores. Learning curves are obtained by plotting the test scores from the CV process, for different sizes of the learning set (incidentally, often both the train and test scores are plotted, so that one can use a single plot to evaluate variance/bias *and* determine whether more data would be helpful). By looking at whether our current learning set size lies in the "slope-y" or "flat" part of the curve, we can gauge the impact of gathering additional data.

Figure 3.7 shows the learning curves for the two models we are considering, the decision trees and the kNN.

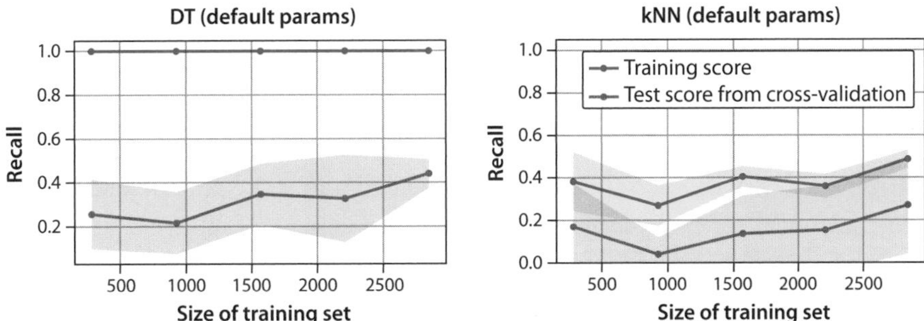


**Figure 3.7: Learning curves for the Decision Tree classifier (left) and kNN classifier (right) with default parameters, plotted on the same y-scale. In both cases, the learning curves suggest that adding more data would be helpful in improving test scores, given the trend with data set size. By showing the train scores as well, we are able to diagnose a strong high variance problem for the DT classifier, revealed by the large gap between train and test scores. In contrast, the kNN classifier suffers from high bias, as indicated by the low train and test scores; the gap in this case is not statistically significant and therefore insufficient for a diagnosis of high variance.**

## 3.6 IMPROVING A SUPERVISED CLASSIFICATION MODEL

After the initial diagnosis phase, we are now ready to take steps to *improve* the performance of our algorithm. We list some ideas here and will go into more detail with their implementation in Chapter 4.

### 3.6.1 Hyperparameter tuning

Hyperparameter tuning is a very important strategy in optimizing ML models, and it is often the first step after basic diagnostics. The hyperparameters define the architecture of the model. For example, in our decision tree, we might decide to limit the maximum number of splits that the classifier is allowed to make, or to set a minimum size of the final leaf nodes so that they are bound to contain at least a certain number of samples. In the kNN algorithm, we might decide to change the number of neighbors (or the radius of the relevant neighborhood, if using the radius-based version) considered in the final average estimate, or move away from uniform weights in the average by assigning more weight to neighbors that are closer to the sample under consideration.

While in general optimizing the parameters of a model is always a good idea, having run some diagnostic tests beforehand can help us determine which parameters are worth varying and choose some possible ranges. In most cases, learning curves that include a train/test scores comparison will point to a high bias or high variance situation. High bias might point to an excessively simple model, while high variance is related to poor generalization, which might stem from using an excessively complex model that has trouble weeding out unnecessary information.

In either case, we might want to use our parameter tuning strategy to effectively increase or decrease *model complexity*. If our model has been diagnosed with high bias, like the kNN model above, reducing the number of neighbors used in the estimate would probably not help, because it would decrease the complexity of the underlying model. In contrast, in our decision tree, which suffers from high variance, we will want to consider hyperparameter tuning strategies that tend to simplify the model, for example by using a finite, small number of splits.

**Parameter or hyperparameter?** Different practitioners often use the two terms interchangeably, but technically we should reserve the term "hyperparameters" for the variables that relate to the architecture of the model, while "parameters" are the numerical variables of each model that are learned during the training process. For example, if we are employing a polynomial relationship model, the degree of the polynomial is a hyperparameter, while the coefficients of each monomial are the parameters; or the size of layers in a neural network is a hyperparameter, while the weights are parameters (e.g., [Probst et al., 2020])

### 3.6.2 Feature engineering and feature selection

Features are the building blocks of ML algorithms, and our choice of features can play a major role in how well algorithms can process the information that we provide; we saw a simple example of this in Chapter 2. ML models don't have any domain knowledge, so they take each feature at face value, as a series of numbers. Even if many algorithms have built-in mechanisms for recognizing hidden patterns, it is always advantageous to present the information in the most effective possible manner. For example, consider the representation of the habitable planets data set shown in Figure 3.1. If we use a decision tree to partition the data set, we can only make splits that are aligned with each feature; in the figure's two-dimensional projection, this corresponds to horizontal or vertical splits. However, it is clear that a more meaningful partition of the data would be along a diagonal line. This suggests that *creating* a new feature that combines the stellar mass and orbital period might make it easier for a tree-based algorithm to achieve a better result with fewer splits. Someone with knowledge of the domain (in this case, basic physics/Earth science) would have been able to build this feature even before data exploration. In fact, the surface temperature of a planet, which can act as a zero-order proxy for habitability, depends on the luminosity of the parent star and its distance, as those determine the energy received by the planet. Since for most stars the luminosity traces the mass, there must exist an appropriate combination of the parent star's mass and its distance that effectively correlates with habitability.

The process of building new features in order to help ML models sift through information more efficiently is known as *feature engineering*. As we have seen, it can happen in various ways: as a consequence of physical intuition, when scientists use their knowledge of a domain to introduce combinations of variables that are expected to be correlated with the target property, or as a result of a data exploration process (e.g., plotting different existing features). Or it can happen as a systematic search (e.g., by creating all polynomial combinations of current features up to a given degree). Feature engineering is especially expected to help *high bias* cases, if the bias originates in the difficulty that the algorithm has in recognizing a complex pattern.

There is also another side of manipulating features, which works in the opposite direction: the process of *feature selection*, where we aim to select a smaller subset of the original features that carry the bulk of the information. In this sense, feature selection can be seen as a form of dimensionality reduction.

The motivation for this streamlining approach can be to build understanding of a complex problem by identifying the most meaningful players, to make our algorithms more manageable by downscaling the data set size, or to help reduce high variance. In the latter case, the rationale is that having features that are highly correlated or very noisy might act as a confusion effect for our ML models (or in some cases, simply having many features!), making it more likely that the model would overfit.

Finally, some algorithms, like Support Vector Machines or neural networks, contain automated feature engineering steps, which makes them naturally efficient and can positively affect performance. We will consider them respectively in Chapters 4 and 8.

### 3.6.3 Switching algorithms

If parameter optimization and feature selection/engineering don't bring the hoped-for improvements, it might be time to switch algorithms altogether. Different algorithms are engineered to model different relationships, so even the "best version" of a particular algorithm might not be suited to solve a certain problem. To return to the case of our sample algorithms, it is easy to imagine that if we have a small learning set, an algorithm like kNN might not perform well, especially on examples that are near the boundaries of the learning set distribution, while parametric algorithms that effectively create a local model of the I/O relationship might work better for this specific set. Similarly, while a Decision Tree algorithm is bound to make splits along the provided features, algorithms that remap the features to a higher dimensional space might perform better. We will see examples of more sophisticated algorithms in the next few chapters.

### 3.6.4 Resampling

A technique that might be useful in classification problems when the data set is heavily imbalanced is resampling the data to obtain a more balanced distribution before deploying a classifier. The resampled data set is obtained by sampling with replacement from the original one, with an increased probability to sample from the less common class; the most common automated pipeline is to sample with probability equal to the inverse of class frequency. This technique serves to effectively weigh one type of error (typically the false negatives, if we are interested in looking for a rare object) more than the other.

### 3.6.5 Collecting more data

Learning curves have the important function of revealing whether more data would help increase the performance of our predictive modes. If all our other optimization attempts have failed to yield the desired performance, and learning curves for our data set size have not yet stabilized, our next step might just be collecting more data.

### 3.6.6 Everything is correlated

Something important to keep in mind is that optimizing an ML model is not always a sequential process. From the first steps in thinking about the problem and collecting knowledge, through data exploration, diagnostics, and optimization, we are bound to encounter issues, change strategy, fight different issues, go back on our steps as we develop new intuitions, and so on. And it's definitely true that one strategy doesn't fit all: We have to use our judgment to find the holistic best solution, which will take into account how much time we might be able to invest in the process, whether transparency and interpretability are an issue, how costly it is to collect data, and so on. Only on paper (or on the exam!) is the best algorithm the one with the best test scores. Nonetheless, I did my best to create a simplified flow chart that summarizes my recommended approach when we want to solve a problem using machine learning, shown in Figure 3.8.

Finally, a reminder that the diagnostics and optimization techniques described in this section, with the possible exception of resampling, apply to *regression* problems as well. While the metrics used to evaluate the success of regression algorithms will differ, as we will see, beginning in Chapter 5, the model building pipeline will (thankfully!) essentially remain the same.

## 3.7 BEYOND BINARY CLASSIFICATION

Our final consideration for this chapter is how the evaluation might change if our problem has more than two possible outputs, in the so-called multiclass case. Such classification problems are very common in science and beyond; for example, face recognition algorithms that tag specific people are running a classification algorithm with as many possible outputs as known subjects.

For multiclass problems, accuracy is still easily interpretable and applicable as the total number of correct classifications. Similarly, we can build the confusion matrix without ambiguities; it will be $n \times n$ (instead of $2\times2$), and we still can assume that the row index indicates the intrinsic (note how we shouldn't say "true," which is used to indicate the correctness of the classification by a model!) class membership, and the column index indicates the predicted class membership.

However, evaluation metrics like precision and recall rely on the concept of a "positive" class, so if we want to employ them, it is necessary to rephrase their definitions in a framework akin to binary classification. The most common approach is the *one-versus-all*. In this formalism, we start by assigning the "positive class" label to one of the classes and consider all remaining elements as members of the negative class. We can then build an "effective" binary classifier, for which any metric can be calculated. We continue this process by cycling through all classes, regarding each of them as the positive class once. This gives us $n$ estimates of metrics for an $n$-class problem; let's assume we are interested in precision for this example's sake, since we will use the explicit definition below, and let's call them $P_1, P_2, \ldots P_n$. The final step is to *average* the performances to obtain one final estimate. We can do it in two ways:

$$P = \frac{1}{n} \sum P_1, \ldots . P_n, \tag{3.1}$$

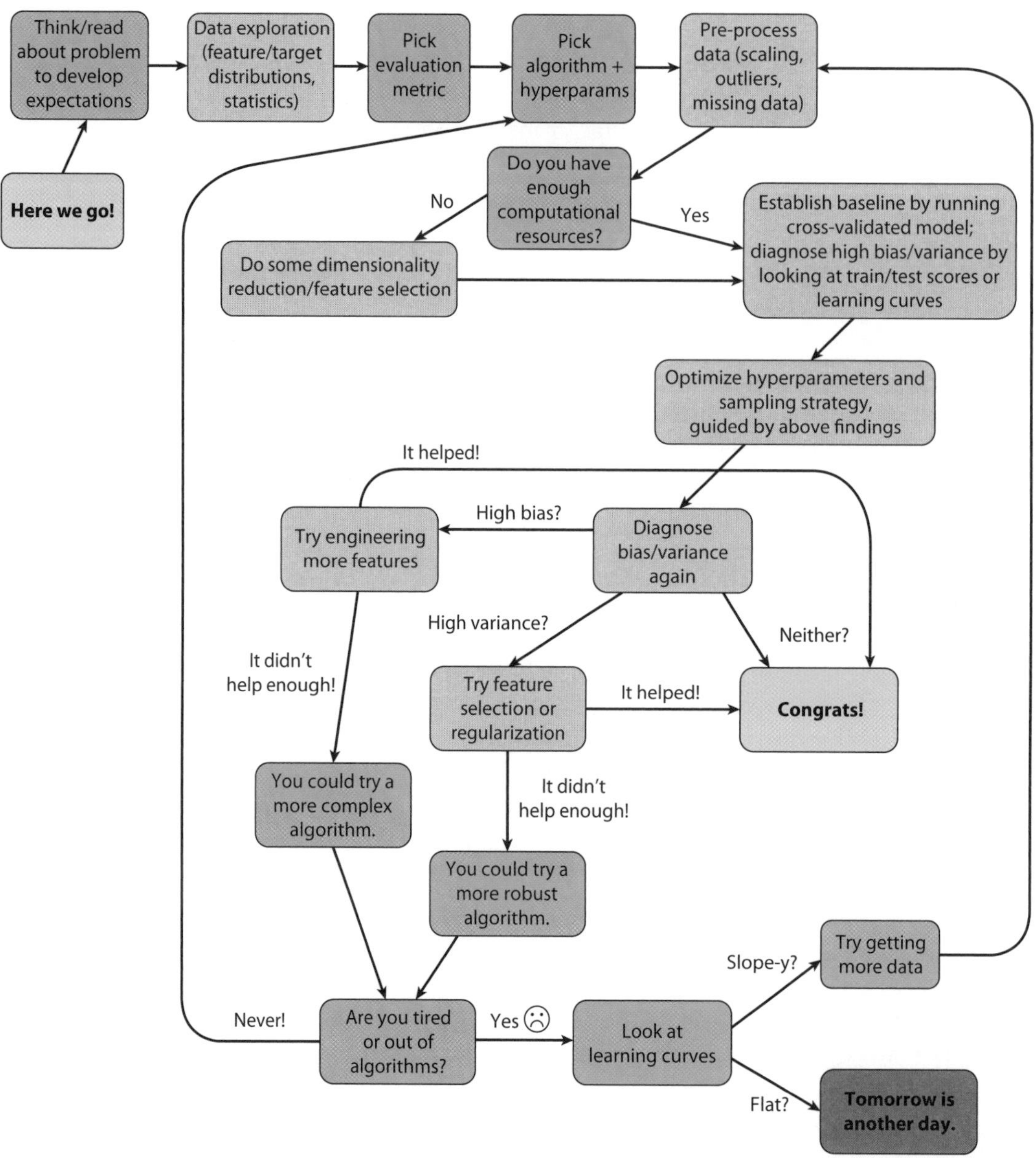


Figure 3.8: **A simplified flow chart showing what the problem-solving process might look like in machine learning. Light gray boxes indicate computational steps; blue boxes indicate "critical thinking" steps.**

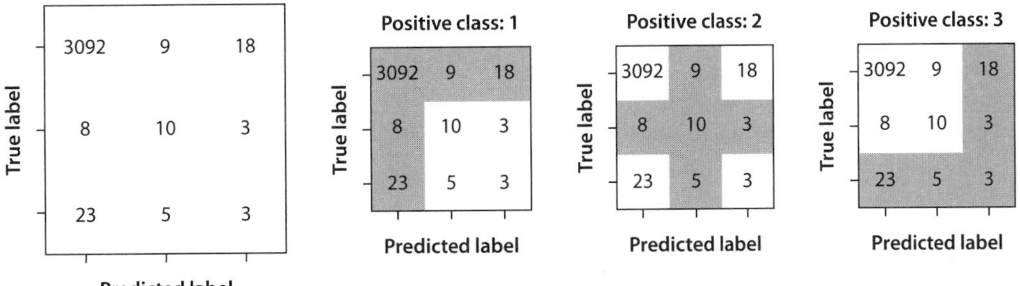


**Figure 3.9: Larger panel: Confusion matrix for the Decision Tree algorithm with default parameters, in the 3-class formulation of this problem. The classes are: not habitable, optimistically expected to be habitable, and reasonably expected to be habitable. Smaller panels: The three "one-vs-all" confusion matrices, useful for calculating class-specific metrics, such as precision and recall. The positive class is indicated by the darker shade.**

that is, an arithmetic mean of the $n$ estimates, independent of class size; this is called *macro averaging*. Alternatively, we can perform the average at the building blocks level; remembering that $P$ is defined as $\frac{TP}{TP+FP}$,

$$P = \frac{TP_1 + TP_2 \cdots + TP_n}{TP_1 + TP_2 \cdots + TP_n + FP_1 + FP_2 \cdots + FP_n}. \tag{3.2}$$

This approach, called *micro averaging*, weighs good performance in high-membership classes more heavily.

We can look at one example that shows the difference between the two approaches by going back to our original data set and remembering that the planet habitability label was not binary but had three possible values: 0 (for not habitable), 1 (for optimistically habitable), and 2 (for reasonably expected to be habitable). By running the Decision Tree algorithm with default parameters and combining the predicted test labels with the `cross_val_predict` function, we obtain the confusion matrix in Figure 3.9, which illustrates the 3-class setting.

The estimates of TP, TN, FP, and FN for each of the three possible iterations of the positive class can be obtained by dividing the matrix in a one (positive) versus all others (negative) approach. In Figure 3.9 we see the three possible matrices, with the positive class indicated by the darker shade in each case. The true positives are found in the shaded box on the negative-sloped diagonal. The true negatives are obtained by adding the numbers in the white cells (note that unlike the binary case, not all true negatives are correct classifications, because swapping objects between negative categories is not counted as an error). The false positives are the sum of off-diagonal terms of the shaded column, and the false negatives are the sum of off-diagonal terms of the shaded row.

Hence,

$$TP_1 = 3092; \quad TN_1 = 21; \quad FP_1 = 31; \quad FN_1 = 27; \tag{3.3}$$

$$TP_2 = 10; \quad TN_2 = 3138; \quad FP_2 = 14; \quad FN_2 = 11; \tag{3.4}$$

$$TP_3 = 3; \quad TN_2 = 3119; \quad FP_3 = 21; \quad FN_3 = 28. \tag{3.5}$$

From these numbers we can easily derive the class-specific precision estimates, which are 0.99, 0.42, and 0.125, respectively; the large number for the first class comes from how easy it is to identify very common objects.

The corresponding macro average, which is insensitive to class size, would be $P_{\text{macro}} = 0.51$. The micro average, which is dominated by the most abundant class, would be $P_{\text{macro}} = 0.98$. You should consider which one is more meaningful in this case.

Finally, note that in `sklearn`, it's very easy to estimate these metrics by making use of the classification report attribute:

```
print(metrics.classification_report(nb_targets, nb_pred)),
```

which will calculate the micro and macro averages of several relevant metrics.

## 3.8 LESSONS LEARNED

As usual, the chapter concludes with a schematic summary of lessons learned and open issues that we will tackle in Chapter 4.

- We always begin by preliminary data exploration, to know our data better and to anticipate potential pitfalls. For example, is the data set very imbalanced? Are there obvious outliers and/or missing data, and if so, how do we want to treat them? Is some preprocessing step needed? Also, spending time engineering sensible features at this stage might be a great investment.
- Next, we think about how we will evaluate our models. There is no universally right metric; we need to be aware of what is important for us (typically, for the science question we are trying to solve), and often this will correspond to defining a custom metric. For a binary classification problem, any metric will be some combination of TPs, TNs, FPs, and FNs.
- Models are almost never perfect at first try, and they tend to suffer from high bias (underfitting, or lack of accuracy) and/or high variance (overfitting, or difficulty to generalize). Diagnostic tools such as comparing training/test scores and learning curves can help us understand where there is room for improvement.
- Hyperparameter tuning, resampling, feature engineering, and feature selection can all be used to improve a classifier's performance. It is very important that any

preprocessing step is carried out while blinding the test data, to avoid leakage of information during the training process.

- A multiclass classification model can be evaluated in a similar manner to a binary classifier by recasting the problem as one-vs-all.

## 3.9 REVIEW AND DISCUSSION QUESTIONS

Note: Questions and exercises marked by ** are more complex, open-ended, or time consuming. Those marked by * have more than one correct answer.

**Exercise 1.** * Why do we do k-fold cross validation rather than using a simple training/test split?

A) To solve underfitting
B) To avoid over/underestimating performance because of a peculiar split
C) To utilize all data for training
D) To select better features

**Exercise 2.** You are dealing with a binary classification problem in an imbalanced data set. Which one of the following should you probably **not** use as an evaluation metric?

A) Accuracy
B) Precision
C) Recall
D) F1 score

**Exercise 3.** You want to analyze the online literature about a rare type of toy hugging bear. You create a classifier that categorizes textual information about such bears and other animals. Which metric should you pick to make sure you collect all the existing information?

A) Accuracy
B) Precision
C) Recall
D) F1 score

**Exercise 4.** You want to select the 10 best gifs of cats of all time. Assuming that the "cat" class is the positive classification, which metric should you maximize in an algorithm that collects potential candidate images?

A) Accuracy
B) Precision
C) Recall
D) F1 score

**Exercise 5.** Your algorithm suffers from high bias if:

A) It does not generalize well.
B) The model you are using is too simplistic.
C) The model you are using is too complex.
D) The algorithm you are using takes too long.

**Exercise 6.** * Your algorithm suffers from high variance if:

A) It does not generalize well.
B) The model you are using is too simplistic.
C) The model you are using is too complex.
D) The algorithm you are using takes too long.

**Exercise 7.** A telltale sign that your algorithm suffers from high bias is:

A) The training scores are much better than the test scores.
B) The training scores and the test scores are poor.
C) Your scores are too good to be true.
D) The learning curves show that the training scores have plateaued.

**Exercise 8.** A telltale sign that your algorithm suffers from high variance is:

A) The training scores are much better than the test scores.
B) The training scores and the test scores are poor.

C) Your scores are too good to be true.
D) The learning curves show that the training scores have plateaued.

**Exercise 9.** * Which of these strategies may help fix high bias?

A) Increasing model complexity
B) Decreasing model complexity
C) Using more or different features
D) Using fewer features

**Exercise 10.** * Which of these strategies may help fix high variance?

A) Increasing model complexity
B) Decreasing model complexity
C) Using more or different features
D) Using fewer features

**Exercise 11.** ** Which type of score averaging, micro or macro, would you recommend using in an imbalanced multi-class classification problem? Why?

## 3.10 PROGRAMMING EXERCISES

**Exercise 1.** For the problem discussed in this chapter, consider a different imputing strategy, for example, using the mean of the column to fill missing values, and repeat the classification exercise with a method of your choice. What has changed? Which strategy would you recommend?

**Exercise 2.** Use one of the models from Chapter 2 (the first/second decision tree or the kNN, trained on the small learning set of 18 objects), without retraining it, to predict labels for the data set of this chapter. What is the model's performance? Was it expected? How could you have known beforehand?

**Exercise 3.** Build a kNN classifier that predicts planet habitability for the data set described in this chapter. Use the `StandardScaler` and the `RobustScaler` methods (both available in the `sklearn.preprocessing` module) to scale the features before using the kNN algorithm. Does this choice make a difference? Which one would you recommend using for this case?

**Exercise 4.** ** Based on your knowledge of the laws of physics (in particular, the scaling of stellar luminosity as a function of mass for main sequence stars and the scaling of flux from a point source with distance), engineer a new feature (a combination of two or more features you already have) that might correlate strongly with planet habitability. Explain why you think this feature would work.

**Exercise 5.** ** Add your new feature to the data frame and rebuild the decision tree model we built in the "HP_Chap3.ipynb" notebook. Do the results improve? Explain why or why not.

**Exercise 6.** ** Consider the planet habitability data set with its original label structure of three possible categories, corresponding to not habitable, optimistically habitable, and reasonably expected to be habitable (shown in Section 3.7).

- Optimize one algorithm of your choice using an appropriate evaluation metric for a 3-class classification problem.
- Print or plot the confusion matrix: What is the most common mistake your classifier is making? Discuss this result.
- The "`classification report`" attribute allows you to calculate the precision and recall scores for each of the 3 classes. Would you recommend macro averaging or micro averaging for this case? Why?

CHAPTER FOUR

# Supervised Learning Models: Optimization

*What do you call an overexcited tunable number?*
*A hyperparameter.*

In this chapter, we will attempt to put together all the lessons learned so far about exploring, diagnosing, and optimizing supervised machine learning (ML) models in order to build a complete pipeline. We will continue using a classification problem as an example, but keep in mind that most of this material, with the exception of evaluation metrics and a few other classification-specific details (e.g., stratification), applies to regression problems as well. The main new piece of content will be the introduction of nested cross validation (CV) as the tool to estimate generalization error appropriately when we are optimizing hyperparameters.

We will use a new data set, supplied by particle physicists working on the Standard Model of particle physics, in which we try to predict the nature of collisional events, based on the observed particle produced in the collisions.

To keep things interesting and learn about new techniques, we will introduce a popular algorithm, *Support Vector Machines* [Boser et al., 1992] or SVMs. It is fundamentally different from the tree-based or distance-based algorithms that we have seen so far, and its mathematics are absolutely fascinating. We will describe the data set and the algorithm in Sections 4.1 and 4.2, and then dive into the problem-solving part beginning in Section 4.3.

## 4.1 DATA SET DESCRIPTION

Particle colliders, such as the Large Hadron Collider (LHC), are designed and built to investigate the quantum interactions between the fundamental building blocks of matter by smashing together particles at very high energies and then studying the outcomes of their collisions.

The enormous amount of data coming from such experiments is provided in the form of collision events (or simply events). Each event is described by the set of

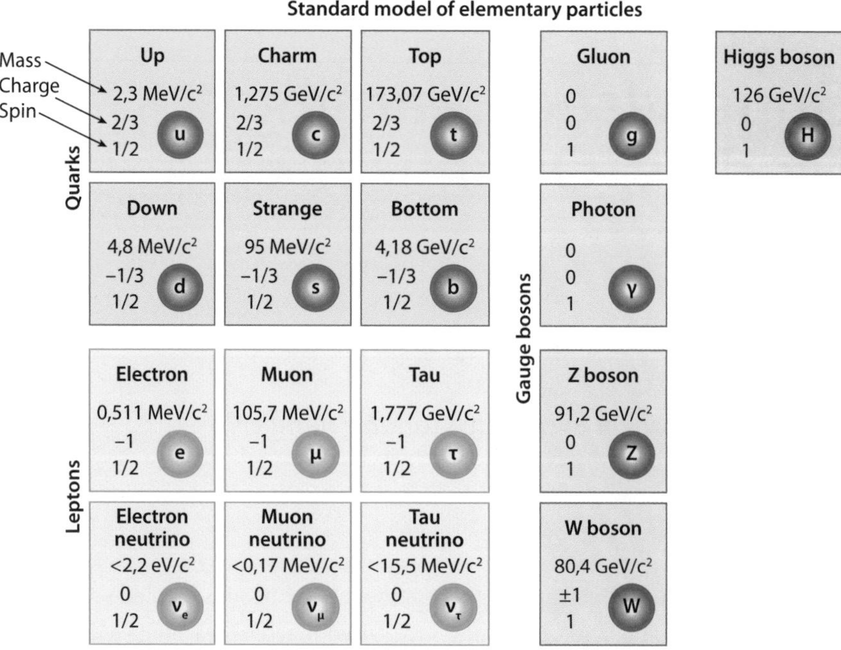


**Figure 4.1: The fundamental particles and families in the Standard Model.**

information that the detectors inside the particle collider were able to record as the aftermath of each individual collision. Typically, this consists of the list of the various particles produced in the collision that reached the detectors and measurements of their physical properties, such as energy and momentum, as well as transverse momentum or the pseudorapidity, which are related to the angular separation of the particle from the direction of collision.

The Standard Model (SM) of particle physics (e.g., [Gaillard et al., 1999]) has been remarkably successful in describing the behavior of particles as observed in particle colliders, most recently at the LHC.

According to the SM, matter is made by fermions: three families of leptons (electrons, muons, tau leptons, and the corresponding neutrinos) and three families of quarks. Protons and neutrons, which form the nucleus of atoms, are examples of hadrons, namely, composite objects made by a combination of different quarks. Leptons and quarks are stable particles and have fairly small masses. The one exception is the top quark, whose mass is above 170 GeV. When produced in particle collisions, the top quark decays almost instantaneously. This fact will be relevant for our example.

Matter particles interact with each other by exchanging vector bosons, the mediators of the fundamental forces: The photon is the mediator of the electromagnetic interaction; the W and Z bosons are the mediators of the weak interaction, responsible for particle decays; the gluons are the mediators of the strong force, which holds the atomic nuclei together. Note that a fourth fundamental force, gravity, is not part of the SM. While theoretically this is a big open question, in practical terms the effects of gravity are negligible in particle interactions in the colliders.

In recent years, the success of the SM theory was cemented by the discovery of the Higgs boson [Aad et al., 2012; Chatrchyan et al., 2012], often dubbed the last missing piece for the theory's experimental confirmation. The Higgs boson, which is a scalar, unlike the other fundamental bosons described above, is not the mediator of a force but plays a different and important role in Nature: The mass of each particle can be defined by its "amount of interaction" with the Higgs boson (see, e.g., [Veltman, 2018; Schwartz, 2018]).

In spite of the great success of the SM, there are several phenomena, such as the origin of neutrino masses, the nature of dark matter, or the dynamics of electroweak symmetry breaking, that cannot be explained within its realm, motivating the search for new physics Beyond the Standard Model (BSM). The task of developing a framework to detect the experimental signature of BSM physics is challenging but extremely important to achieve any further progress in our understanding of the Universe. Viable investigation strategies include focusing on the predictions of some specific extensions of the SM (e.g., supersymmetry), adding new particles to the Lagrangian of the SM and predicting their interactions with known particles, or adding new interaction channels for existing particles [Brooijmans et al., 2020]. In most cases, the clearest sign of BSM phenomenology would come from noticing a statistically significant deviation between the predicted (by the SM) and observed behavior of particles produced in collisional events.

In our example, we use a simulated data set that focuses on the production of four top quarks in a single event. A signal for the production of these events has recently been observed, with a significance of $4.6\sigma$, by the ATLAS experiment [ATLAS Collaboration, 2021].

The discovery of "4-top events" is interesting, because measuring more events than expected could point to BSM physics. Unfortunately, we can't observe the top quarks directly, as they quickly decay into other particles, which are then measured in the detector surrounding the point where the quarks were originally produced. This is reflected, in the practical terms of our data set, in the absence of top quarks from the list of particles in each event.

Top quarks decay semi-leptonically, as depicted in Figure 4.2, by producing a bottom quark and a W boson. The presence of a bottom quark is observed and recorded in the detector in the form of a cascade of particles known as a b-jet. There is no sign of W bosons in the data set, since they will further decay before reaching the detector. There are two possible options for W-boson decays: hadronic decay into a pair of light quarks, which reach the detector as jets, or leptonic decay into a lepton plus neutrino, which will be recorded in the data set as a lepton (electron/positron or muon) plus missing energy (MET), because the neutrinos escape the detectors.

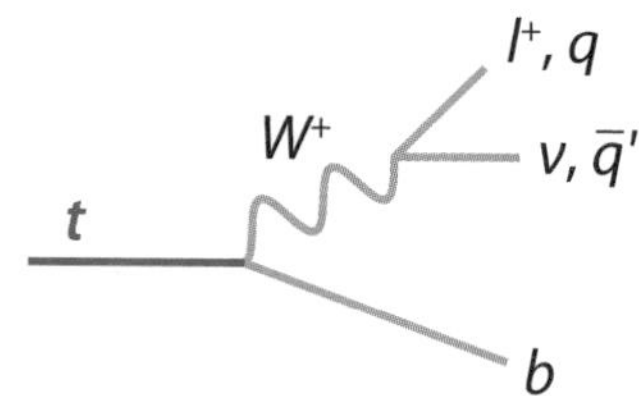


**Figure 4.2: A top quark decays into a W boson and a bottom quark, and the W boson further decays into leptons, neutrinos, and jets. From [Greif and Lannon, 2020].**

The simulated learning data contain 5,000 examples, selected at random from a larger data set of 100,000. Each line contains information about a collision event, in which a varying number of particles are produced. For each particle produced in the event, we know its type: electron (e−), positron (e+), photon (g), a so-called jet (j), a so-called b-jet (b), or a muon (m+/m−) with positive or negative charge. We also know its 4-vector, that is, the energy, transverse component of the momentum, and the $\theta$ (given here in units of pseudorapidity) and $\phi$ angles. The latter quantities are related to the separation of the emitted particle from the axis of the collision, which is important for the selection of clean events. Finally, conservation of energy requires that we account for missing energy—in other words, energy that was carried out by undetected particles, for example, neutrinos. This information is encoded in the "MET" and "METphi" variables, which contain the magnitude and the azimuthal angle of the missing transverse energy vector of the event, respectively.

Overall, the format of the information for each event is the following:

MET; METphi; obj1, E1, pt1, eta1, phi1; obj2, E2, pt2, eta2, phi2; ...

where "obj" specifies the type for each particle detected in the event. Because this is a simulated data set, we also have information about the target variable, the process that generated the event, which can be a "4-top" or "t-tbar" tag. Our goal is to build a classifier that can predict this label correctly, on the basis of the information about the particles produced in the event.

## 4.2 A NEW ALGORITHM: SUPPORT VECTOR MACHINES

Support Vector Machines (SVMs) are a popular supervised learning algorithm suitable for classification or regression; in this chapter, we will apply it to the binary classification problem discussed in the previous section. In a nutshell, the algorithm works by searching the hyperplane that provides optimal separation between the two classes; in a multiclass problem, this will be done using a one-vs-all approach.

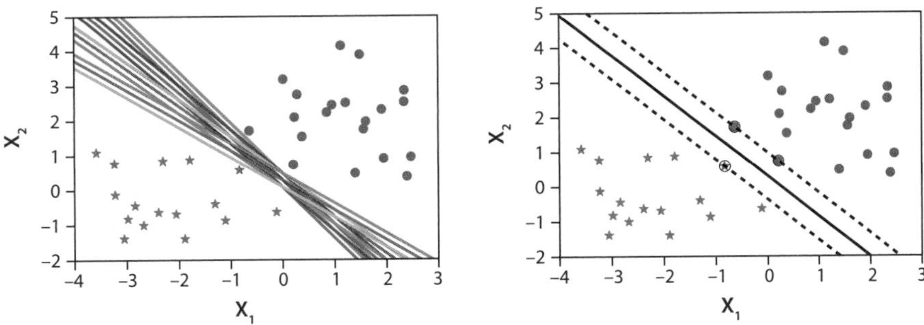


**Figure 4.3: This simple data set can be linearly separated in a myriad of different ways (left), but only one choice maximizes the width of the margin between the two classes (right).**

## 4.2.1 Linearly separable data

Let us illustrate the idea by looking at the simplest possible case: a linearly separable data set, such as the one shown in Figure 4.3. In this example, the two classes can be separated perfectly by drawing a line (which is a one-dimensional hyperplane). However, as shown in the left panel of the figure, there are many lines that can do the job! Which one should we choose?

Intuitively, we know that a line that goes very close to some of the training examples might be a risky choice; we would want the two classes to be as widely separated as possible in order to increase our confidence in the predicted classifications. We can formalize this requirement by introducing the idea of a margin. Consider any possible separation between the two classes, and pick for each one the closest example(s) from the positive and negative class. The *margin* of an SVM is defined as the empty space described by the two parallel lines that touch these closest examples, and the *decision boundary* of the SVM will be the line in the middle of the margin (so that the decision boundary is equidistant from the two classes). The optimal boundary is defined as the one that *maximizes the width of the margin*, achieving maximal separation. The examples from each class that determine the decision boundary are called the *support vectors*; their number depends on the spatial distribution of the training examples in feature space, with a minimum of two (e.g., [Bennett and Bredensteiner, 2010]). The right panel of Figure 4.3 shows the optimal separator for our example; in this case, there are three support vectors, which have been circled with a black line in the figure. Because the decision boundary is uniquely determined by the support vectors, the classifier's decision is only influenced by them, as opposed to by the entire training set; this helps curb the computational cost of the SVM even for a large number of features or examples.

It is useful to introduce a bit of mathematical notation before proceeding further, which will help show how the SVM algorithm can be generalized to nonseparable data sets and nonlinear boundaries. The mathematics of SVMs is described in great detail in several references; here we adopt the formalism and general philosophy of [Ben-Hur and Weston, 2010] and [Ng, 2019].

A linear classifier, such as the one we have been considering so far, has a decision function that is linear in the input examples and can be expressed as

$$f(\mathbf{x}) = \mathbf{w}^T\mathbf{x} + b. \tag{4.1}$$

The vector $\mathbf{w}$ is known as the weight vector; $b$ is called the bias, and $\mathbf{w}^T\mathbf{x}$ is the dot product between the weight vector and any input example $\mathbf{x}$. The decision boundary of this classifier is defined by the equation

$$\mathbf{w}^T\mathbf{x} + b = 0; \tag{4.2}$$

in other words, it is where the decision function vanishes.

The hyperplane defined by this equation divides the space into two according to the sign of the discriminant function $f(x)$. We can always redefine the labels so that the positive examples are characterized by $f(x) > 0$, and the negative examples are characterized by $f(x) < 0$.

We can now express the width of the margin in terms of the weight vector of a linear classifier. The decision function for the closest examples from each class ($x_+$ and $x_-$) will be

$$\begin{aligned} f(\mathbf{x}_+) &= \mathbf{w}^T\mathbf{x}_+ + b = a > 0, \\ f(\mathbf{x}_-) &= \mathbf{w}^T\mathbf{x}_- + b = -a > 0, \end{aligned} \tag{4.3}$$

where $a$ is the distance from the decision boundary, which is assumed to be equidistant from $x_+$ and $x_-$. We can rescale the $(\mathbf{w}, b)$ coordinates of our system by a factor $1/a$ to obtain the equation of the hyperplanes (lines, in our case), defining the margin as

$$\begin{aligned} f(\mathbf{x}_+) &= \mathbf{w}^T\mathbf{x}_+ + b = 1, \\ f(\mathbf{x}_-) &= \mathbf{w}^T\mathbf{x}_- + b = -1. \end{aligned} \tag{4.4}$$

In this new coordinate system, the decision function for examples outside the margin will be $f(x) > 1$ or $f(x) < -1$.

From the above equations we can derive the width of the margin as $\mathrm{d} = 2/||w||$; that is, the width of the margin is inversely proportional to the norm of the weight vector.

Finding the optimal classifier can now be expressed as a constrained optimization problem: We want to maximize the width of the margin subject to the condition that the margin is empty (no input examples are allowed within the margin). In mathematical terms, our problem can be written as

$$\begin{aligned} &\text{minimize}_{w,b} \quad \frac{1}{2}||w||^2 \\ &\text{subject to} \quad y^i(\mathbf{w}^T\mathbf{x_i} + b) \geq 1, \end{aligned} \tag{4.5}$$

where $\mathbf{x}_i$ are the training examples, the first condition stems from the fact that maximizing the width of the margin $(2/||w||)$ is equivalent to minimizing $\frac{1}{2}||w||^2$, and the second condition uses the fact that $y^i = -1$ for negative examples and $y^i = +1$ for positive examples.

We will not go into the details of how the optimization problem can be solved, but one of the most notable results is that the weight vector can be expressed as a linear combination of the training examples $\mathbf{x}_i$:

$$\mathbf{w} = \sum_{i=1}^{n} y_i \alpha_i \mathbf{x}_i. \tag{4.6}$$

The examples for which $\alpha_i$ are nonzero are the support vectors, so that the classifier's decision $f(\mathbf{x})$ only depends on the dot product of the example under consideration, $\mathbf{x}$, with the support vectors, as mentioned in the previous section. The expansion in terms of the support vectors is often sparse; the fraction of the data serving as support vectors is an upper bound on the error rate of the classifier [Schölkopf and Smola, 2001].

### 4.2.2 Nonseparable data sets and slack variables

The simple margin maximization criterion described above loses its formal applicability for data sets that are not linearly separable. However, we can imagine many cases in which a linear separation between classes may still be a reasonable choice, barring some noisy examples. As an example, let's look at the left panel of Figure 4.4. This data set cannot be *exactly* separated with a linear boundary; however, a line still provides a reasonably well-behaved separation between the classes, and it would take a much more complex and "wiggly" decision boundary to be able to pull off an exact separation. Increasing model complexity might result in high variance, as we have learned; so we might be better off keeping a simple linear model, even at the price of some incorrect classifications on the training set.

How can we apply the concept of margin to not exactly separable cases? Our optimization criterion had been very simple: maximizing the width of the margin (in other words, our loss function was simply $1/||w||$). Here, we might have to accept that (1) There will be examples that lie within the margin ("on the street," as it is said sometimes in SVM jargon); and (2) There will be examples that are misclassified. We can attribute a penalty to each of these cases by introducing "slack variables" $\xi_i$ that describe the distance between the margin and the examples. Intuitively, it's worse to misclassify an example very far from the margin, compared to

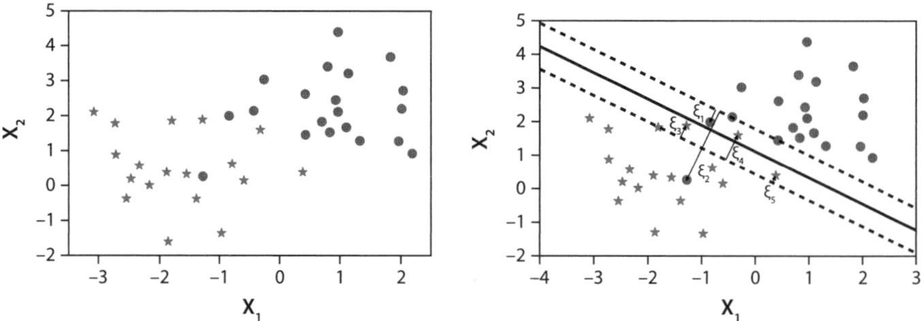

Figure 4.4: This data set cannot be exactly separated by a line (left), but we can obtain a reasonably well-behaved classifier by allowing some examples to lie within the margin and some misclassifications to be near the boundary. These are assigned a penalty in the loss function that is proportional to the distance $\xi_i$ to the margin of the corresponding class (right).

an example close to or within the margin, as we always expect the most uncertain or difficult cases to lie close to the decision boundary. Therefore, we can attempt to maximize the width of the margin (or, equivalently, minimize the L2 norm of the weight vector) while we minimize the sum of the penalty terms associated with examples that are within the margin (for which $0 < \xi_i < 1$) or are misclassified (for which $\xi_i > 1$). The margin is in this case a *soft margin,* in contrast with the margin of the exactly separable case, which is a *hard margin.* Figure 4.4 shows an example of a data set that is not exactly separable and the soft margin solution found by a linear classifier.

In terms of a mathematical formulation, the problem can be recast as

$$\begin{aligned} &\text{minimize}_{w,b} \quad \frac{1}{2}\|w\|^2 + C\sum_i \xi_i \\ &\text{subject to} \quad y^i(\mathbf{w}^T\mathbf{x_i} + b) \geq 1, \end{aligned} \tag{4.7}$$

where $C$ is a regularization parameter that regulates the trade-off between having a wide margin and having few misclassifications/within-margin examples; its behavior is further illustrated in the next subsection.

### 4.2.3 Nonlinearly separable data and kernel functions

Now that we have learned how to treat data that are linearly separable with some noise, we can turn to the more realistic cases of nonlinearly separable examples. The powerful idea behind the SVM algorithm is that by an appropriate coordinate (i.e., feature) transformation, we can project the data onto a higher-dimensional space where they can be linearly separated (with some noise). This allows us to recycle the simple and efficient machinery of linear classifiers and use them in the transformed feature space.

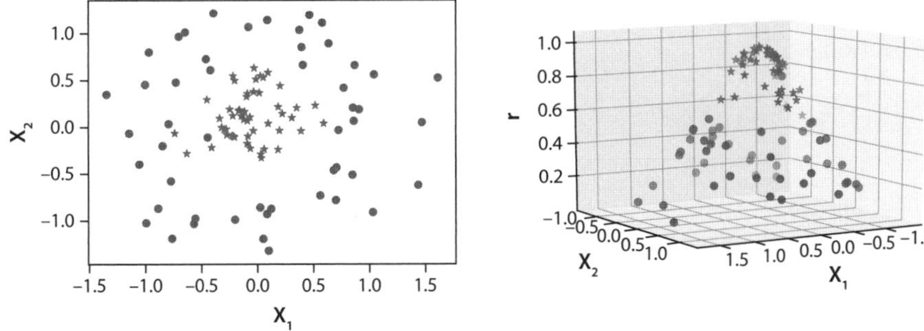


**Figure 4.5: This data set cannot be separated by a linear classifier (left), but we can turn the problem into a linearly separable one by mapping the points onto a three-dimensional space (right).**

An example of such a coordinate transformation is given in Figure 4.5. In the left panel, we see some two-dimensional data, which are clearly nonlinearly separable; however, we can see that classes could be separated with an ellipse-like boundary. The SVM machinery handles this problem in feature space: If we use a Gaussian transformation to create a third (azimuthal) coordinate based on the distance from the center of the distribution, we can map the data to a three-dimensional space in which the two classes can be separated (with some noise) by a horizontal hyperplane.

This example shows that if we can elaborate an appropriate coordinate (feature) transformation from the original space with feature vector $\mathbf{x}$ to a higher-dimensional space with feature vector $\phi(\mathbf{x})$, we can obtain the decision function in terms of the dot products between $\phi(\mathbf{x})$ and $\phi(\mathbf{z})$, where $\phi(\mathbf{z})$ are the support vectors.

The question that naturally arises is: How can we find the $\phi(\mathbf{x})$ transformation? And here the "magic" of kernel SVMs arises. It can be shown that for some specific functions $K(\mathbf{x}, \mathbf{z})$, known as *Mercer kernels*, any possible feature mapping involving the dot product between $\phi(\mathbf{x})$ and $\phi(\mathbf{z})$ can be replaced by an appropriate kernel function $K(\mathbf{x}, \mathbf{z})$; and equivalently, any valid Mercer kernel corresponds to a specific feature mapping.

The practical advantage of this so-called kernel trick is two-fold: one, we can explore different feature mappings through different kernel functions, and two, the complexity of calculating the kernel function $K(\mathbf{x}, \mathbf{z})$ is $\mathcal{O}(N)$, even when the dimensionality of the implied feature space is much higher. Therefore, the kernel trick provides a powerful and efficient strategy to tackle nonlinear data sets.

### 4.2.4 Hyperparameters

Linear and kernel-based SVMs tend to be one of the most powerful and accurate supervised learning methods before "going neural," but it is really important to tune their hyperparameters appropriately. In this section, we take a look at the effect of varying each parameter and we discuss the appropriate implementation strategy in Section 4.5.3.

#### 4.2.4.1 Type of kernel

The three most common kernel functions (or families) used by SVMs are the linear kernel, the polynomial kernel, and the Gaussian kernel (dubbed "`rbf`" in `sklearn`, for "radial basis function").

The linear kernel corresponds to the simplest case, in which the possible separating surfaces are hyperplanes in the original feature space (i.e., the kernel is effectively the identity matrix).

In the polynomial kernel case, the kernel function has the form

$$K(\mathbf{x}, \mathbf{x}') = (\mathbf{x}^T\mathbf{x} + 1)^d, \tag{4.8}$$

and the SVM works in a new space whose features are all the possible combinations of monomials of degree up to $d$.[1]

Finally, the Gaussian kernel function can be written as

$$K(\mathbf{x}, \mathbf{x}') = e^{-\gamma\|\mathbf{x}-\mathbf{x}'\|^2}. \tag{4.9}$$

The feature mapping corresponding to the Gaussian kernel function is effectively infinite dimensional (but remember that thanks to the kernel trick, we never explicitly compute dot products in the remapped feature space, so the dimensionality is not an issue). There are many clever ways to see that this is the case, but the one that I find easiest to understand is that a Gaussian function is an infinite (converging) Taylor series of polynomials, so one can view the Gaussian kernel as an infinite series of polynomial kernels.

#### 4.2.4.2 Regularization parameter C

The parameter C is the penalty attributed to training examples that are either in the margin or misclassified. As can be seen from the loss function of a soft-margin classifier, Eq. 4.7, a small penalty will yield a wider margin, which is desirable, because it yields a robust separation between classes, while a large penalty will correspond to fewer misclassifications, at the price of a narrower margin. The effect of varying C is illustrated in Figure 4.6 for the linear kernel, but its behavior is common to all kernel types.

For the case of heavily imbalanced data sets, it might be convenient to use a *class-dependent* regularization parameter, whose effect is to attribute a different penalty value to in-margin or misclassified examples for the two classes. This treatment is *de facto* a practical implementation of the resampling strategy described in Section 3.6.4.

---

1 In `sklearn`, the polynomial kernel is based on a slightly different formula, reported at the end of Section 4.2.4.3.

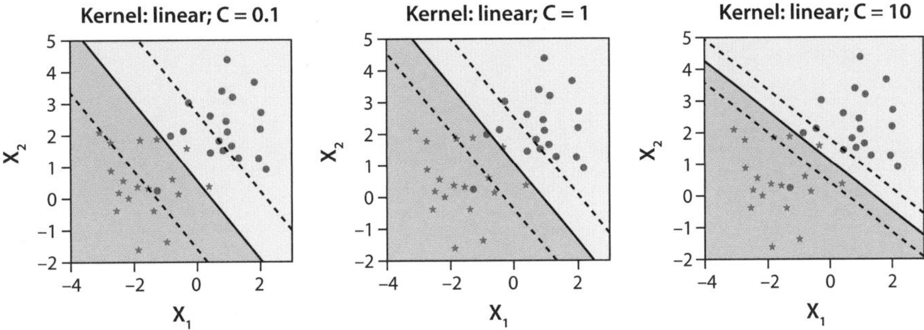


**Figure 4.6: Effect of the regularization parameter *C*. Small values of *C* attribute a small penalty to misclassifications or in-margin examples and yield wider margins. Larger values of *C* attribute a large penalty to misclassifications or in-margin examples and yield narrower margins.**

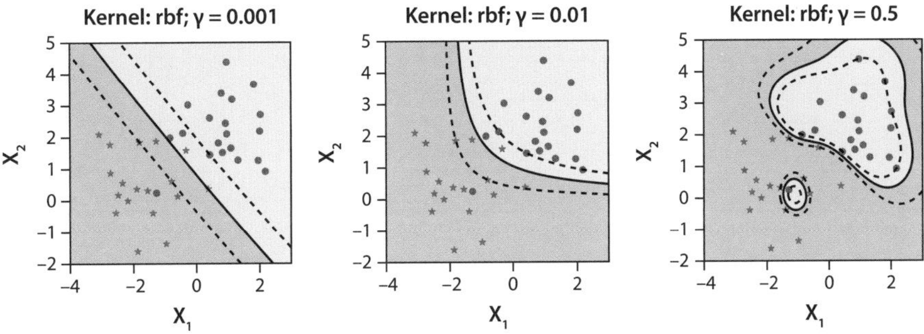


**Figure 4.7: Effect of the $\gamma$ parameter for a Gaussian kernel. Very small values of $\gamma$ are equivalent to a linear boundary; higher values of $\gamma$ generate increasingly complex boundaries.**

#### 4.2.4.3 Gamma parameter

The parameter $\gamma$ defines the "reach radius" of each support vector in the classifier's decision. It is traditionally defined as part of the Gaussian kernel: from Eq. 4.9, we see that support vectors within a radius $1\sqrt{\gamma}$ affect the decision for an example with coordinates $\mathbf{x}$. If $\gamma$ is very small, all support vectors participate in the decision; this generates a smooth boundary (and a linear one in the limit $\gamma \to 0$). If $\gamma$ is large, the decision has a higher degree of locality; this corresponds to increasingly feature-rich ("wiggly") boundaries, similarly to what happens when increasing the degree parameter of a polynomial kernel, as discussed in the next subsection. Note that the polynomial kernel in `sklearn` also uses a $\gamma$ parameter; the kernel function is $(\gamma\mathbf{x}^T\mathbf{x} + c)^d$.

#### 4.2.4.4 Degree parameter

For the polynomial kernel described by Eq. 4.8, if the original space has dimension $n$, the new space has dimension $\binom{n+d}{d}$. The degree parameter $d$ specifies the complexity of the resulting boundary; the linear kernel is a special case corresponding to

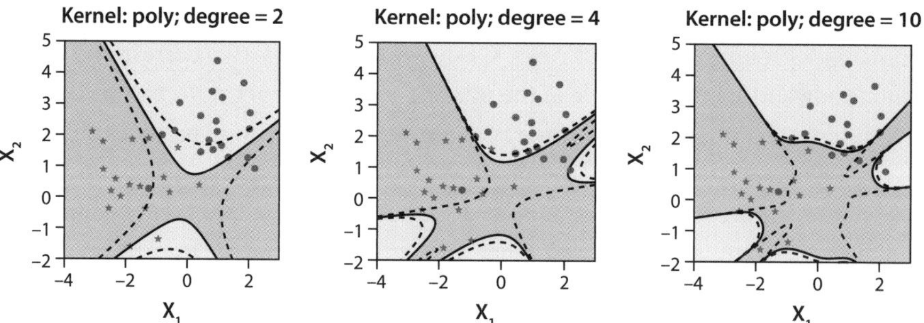


**Figure 4.8: Effect of changing the degree parameter for a polynomial kernel. Smaller values correspond to simpler boundaries, higher values to more feature-rich boundaries.**

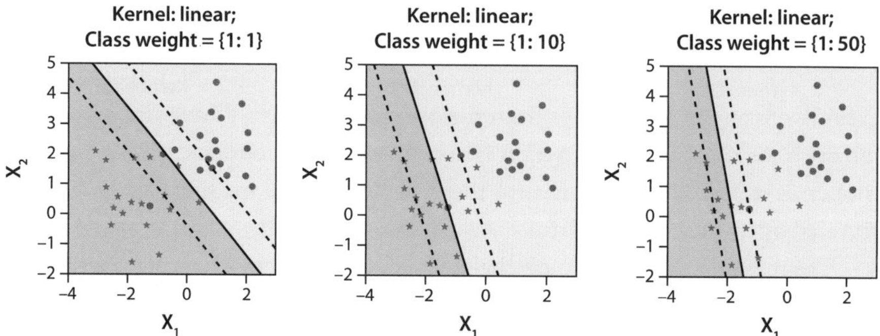


**Figure 4.9: Effect of changing the class weight parameter for a linear kernel. Using a class weight distribution different from 1:1 attributes a different penalty to misclassifications in one class, pushing the boundary toward one or the other set of examples.**

$d = 1$. The effect of increasing the degree parameter is shown in Figure 4.8; higher values correspond to increasing model complexity.

#### 4.2.4.5 Class weight parameter

Finally, a parameter that is handy to vary (especially in the presence of imbalanced data sets) is the *class weight*. Using a class weight distribution different from 1:1 attributes a different penalty to misclassifications in one class and can be used to effectively set a preference for one type of error (e.g., false positives) versus the other. The practical effect of varying this parameter is to ensure a preference for optimizing precision at the expense of recall, or the other way around. We show this effect in the middle and right panels of Figure 4.9. As we increase the class weight ratio, the boundary and margin shift to eliminate any classification error in the "circle" class, while accepting additional classification errors in the "star" class.

## 4.3 DATA EXPLORATION AND PREPROCESSING

We are now finally ready to take a look at our data set and begin our data exploration journey; you can follow along in the "ParticleID_wSVMs.ipynb" notebook.

We can again read the data in a data frame using `pandas`. The data frame contains 68 columns and 5000 rows. Of those 68 columns, two correspond to the magnitude and azimuthal angle of the missing transverse energy vector; 65 contain five descriptors (object type and energy quadrivector) for up to 13 particles observed in the event, and the last one (the process ID) is the label of the event, which tells us whether we are looking at a 4-top or t-tbar event. This column will become the target we use to train our SVM classifier. We can drop it from the feature data frame and use the `LabelEncoder` utility function to convert it to numerical form (0/1):

```
from sklearn.preprocessing import LabelEncoder
y = LabelEncoder().fit_transform(df.processID)
```

Overall, the number of features in the data set is 67. In `pandas`, we can use the function `describe` to take a look at some summary statistics for each column, noting that this function will automatically include only numerical columns and exclude missing or NaN values from the overall count. As we have already seen in the previous chapters, this function is useful for spotting outliers, whose presence is often revealed by the skewness of the distribution, and missing data, which are revealed by the `count` property.

The main data preprocessing challenge of our data set is that each event has a variable number of particles that are produced and therefore a variable number of available features. For example, we can see that out of a total of 5,000 events, the 4-vector of the first product particle is reported in all 5,000 cases, but the number reduces to 4,997, 4,950, 4,717, 4,002, and 2,817 for the second, third, fourth, fifth, and sixth product particles, respectively (and we only have information about the 13th product particle for 56 cases). This behavior is a feature of this specific data set, but such situations (and in fact, even messier ones) are common when dealing with experimental data.

Thus we will need to decide how to handle those issues. When the problem only concerns a few objects, one possible choice is to discard objects with missing data (or extreme outliers), as we did in Chapter 3 for our habitable planets data set. However, in this case, this approach has two problems: (1) we would need to throw away almost the entire data set; and (2) discarding events in which fewer particles are produced (i.e., those with a highest fraction of missing features) is a *biased* selection strategy, which makes our training data statistically different from, say, any new set of events. If the number of total produced particles correlates significantly with our output (4-top or t-tbar), it would lead to an unfair estimate of generalization error. We discuss some better alternatives in the following section.

### 4.3.1 Imputing

An option that may be preferable to discarding large amounts of data is to come up with an *imputing* strategy for the missing features; in other words, some smart way to replace them so that all instances have the same number of features.

What makes a good imputing strategy? It really depends on what our data look like. If the missing values are few in comparison to the data set size, the impact of our choice would be limited. The main purpose of imputing here would be to help the algorithm run smoothly (i.e., not encounter numerical problems), and our main goal should be that imputed values are not "interesting" enough to actively participate in the decision-making process. In other words, we'd like our trained model to ignore them. The most common imputing strategy is to set the missing features to some fixed value, for example 0 (typically a bad choice) or the mean or median value of that feature, which might convey the message of "nothing special is happening here" better. Note that if we decide to use an imputing strategy that makes use of statistics (e.g., the mean or median), we should only calculate the statistics on the training data, to avoid leakage of information between train and test set.

The situation is trickier when we attempt to use features with a significant fraction of missing values. For example, in the case of our current data set, only a minority of examples have data for more than 10 particles produced in the event. If we don't want to discard those examples, we'll have a more difficult imputing problem. In fact, whenever we choose to input a constant value (e.g., the mean or median of observed values) in a large fraction of samples in a column, the information contained in that column will be "washed out," because the intrinsic variability (variance) of the corresponding feature will be artificially pushed toward zero.

A more sophisticated approach involves using regression techniques to *model* the missing values (e.g., [Little and Rubin, 1986]). Popular implementations include KnnImputer [Troyanskaya et al., 2001], MICE (Multivariate Imputation by Chained Equations) [van Buuren and Groothuis-Oudshoorn, 2011], and MissForest [Stekhoven and Bühlmann, 2012]. The basic scheme is the following:

- Select the feature ("y") for which we want to impute missing values;
- Use the subset of examples for which y is known to train a regression model predicting y from the other features;
- Repeat iteratively, choosing a different feature with missing values as y.

Besides the references cited above, I also found Andrew Gelman's notes[2] to be a great introduction to imputing strategies. In `sklearn`, this type of imputation strategy is supported by the `IterativeImputer` module, which can be used to implement several different regressors as a modeling tool, effectively replicating several of the algorithms mentioned above. Some very recent work proposed that imputing and training models at the same time, rather than sequentially, may

2 http://www.stat.columbia.edu/~gelman/arm/missing.pdf

be beneficial [Morvan et al., 2021]; we can expect rapid developments from this particular area of research.

Whatever imputing strategy we may choose, it is important to include it in our pre-processing pipeline, and to avoid leakage of information between train and test sets.

### 4.3.2 Feature selection and preprocessing

For the purpose of this chapter, we will start by selecting a subset of features for which only limited data imputing is required. For example, we can look at the 4-vectors of only the first four particles (features P1-P16), so that we have at the most 283 missing values (for features P13-P16), or just over 5% of the data set. This choice is obviously arbitrary, and you are invited to consider other imputing strategies in the exercises.

Because our feature selection strategy does not make use of statistical properties of the sample and does not peek at the labels, we don't need to worry about doing this within a train/test pipeline.

Ignoring for the moment the object type corresponding to each product particle, we can build a subset of features that we will use, and replace the missing values with 0:

```
features_lim = features[['MET', 'METphi', 'P1', 'P2', 'P3', 'P4', 'P5', 'P6',
'P7', 'P8', 'P9', 'P10', 'P11', 'P12', 'P13', 'P14', 'P15', 'P16']]
features_lim = features_lim.fillna(0)
```

The success of our operation can be tested by using the `describe` property again, which now shows a total length of 5,000 for every column in our new `features_lim` data frame.

At this time, we should also consider whether scaling our data set's features is relevant and recommended for the algorithm we intend to use. SVMs are, in fact, very sensitive to data scaling, because they make decisions based on distances, and the intrinsic scales of different features are combined (like, e.g., $k$ Nearest Neighbors, and unlike tree-based methods). We will verify this in Section 4.3.4.

### 4.3.3 Target exploration and metric

Finally, before we delve into applying the SVM algorithm to this problem, we can take a look at the distribution of the labels, which helps us anticipate possible issues and pick a metric, and establish a benchmark for our expectations. If we impose that the most interesting class (i.e., the "4-top" event) is the positive one, a quick count of the percentage of "1" values

```
np.sum(target)/len(target)
```

yields 0.1622; in other words, $\sim$16% of instances correspond to "4-top" events. This percentage is not skewed enough to be particularly worrisome, and for the time being, we will use accuracy as our evaluation metric. However, one important piece of information to keep in mind is that a "random" classifier that predicts a class just according to class distribution would have an accuracy of just below 73% ($0.8378^2 + 0.1622^2$), and a classifier that places every object in the negative class would have an accuracy of about 84%. This is relevant because in a different context, those numbers might correspond to a pretty decent performance, while in this case, we aim to improve over this baseline. This is another useful reminder that what constitutes a good metric, and how we judge a "good" performance, is extremely dependent on the problem and data set under consideration.

### 4.3.4 Benchmarking

Our first step will be to choose a "reference" model, typically very simple, which acts as our benchmarking step. In our problem-solving pipeline, this step often corresponds to running a model with default parameters. Because SVMs are a heavily parametric algorithm, we will be more explicit here and start from a *linear* kernel model with *no regularization* (i.e., a very high value for the regularization parameter *C*, which is equivalent to a hard margin SVM). The SVM algorithm in `sklearn` is available for classification (SVC) and regression (SVR); we use the SVC for the problem discussed in this chapter.

```
model = LinearSVC(dual = False, C = 10^5)
```

The "dual" argument is set to False, indicating that we will attempt to solve the primal problem, which increases the efficiency of this particular classifier when $n_{samples} > n_{features}$.

Using the `cross_validate` function with a 5-fold stratified CV yields an estimated accuracy value of $0.831 \pm 0.002$, which is pretty low on the basis of the considerations discussed in the previous section.

However, we did not scale our features, and we know that we should! Scaling should happen *after* picking a train/test split inside the CV procedure, so we can make use of the `pipeline` feature to create a sequence of operations to be carried out for each CV iteration, and build our reference model as follows:

```
model = make_pipeline(StandardScaler(), LinearSVC(dual = False,
C = 10^5))
```

After scaling, the `cross_validate` function with a 5-fold stratified CV yields a much improved estimated accuracy value of $0.893 \pm 0.004$, a significant improvement over the performance of a random or constant-output classifier.

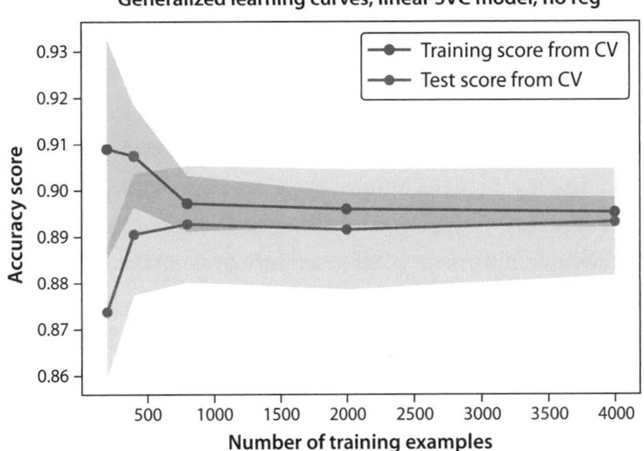


**Figure 4.10: Learning curves for the data set including 18 features, calculated from 5-fold CV for a Linear Support Vector Classifier with no regularization ($C = 10^5$).**

## 4.4 DIAGNOSIS

Once we have "benchmarked" the SVM's performance through the CV process, we can attempt to diagnose the behavior of the algorithm and decide what steps to take in order to further improve its accuracy. One useful "capture it all" strategy is to plot the learning curves showing the evolution of *both training and test set performance*, as a function of data set size.

Learning curves are traditionally meant to evaluate whether an increase in data set size would result in a significant increase in performance, but as we already noticed in Chapter 3, plotting them also gives us the opportunity to diagnose whether we are in a high bias or high variance situation, based on whether there is a significant gap between train and test scores. For this purpose, it is useful to add a measure of the spread in the scores (often simply the standard deviation of the $k$ scores derived from a $k$-fold CV), so we can judge whether any difference is statistically significant.

It is important to note that learning curves are only valid for a specific model and a specific set of parameters; they might change for a different configuration. For this particular (18-feature) data set and for the reference linear SVM model with no regularization, the learning curves are shown in Figure 4.10.

The curves show training/test scores, obtained from 5-fold CV, for several partitions of the data set, corresponding to 5/10/20/50, and 100% of the data, respectively (note that the x axis shows the size of the training set, so the highest value is actually 4,000, or 80% of the full data set).

The training scores start high and then decrease and flatten; this is quite typical, because it's easier to get a great performance on a small training set. The test scores tend to increase as a function of data set size and then flatten. From this graph, we can conclude that *for this particular algorithm and hyperparameter values,* (1) adding more data would not help, and (2) our model does not suffer from high variance, because there is no significant difference between train and test scores. So any steps toward improving the model should go in the direction of solving high bias, for example, by increasing model complexity or by engineering additional features. We can begin our journey by optimizing the parameters of the SVM; in particular, we will aim to include more complex kernel functions that can help our algorithm model the input/output relationship more accurately.

## 4.5 HYPERPARAMETER OPTIMIZATION

Hyperparameter optimization is an almost obligatory step in solving problems with ML. There are two separate sides to this issue: one, how to select optimal parameters; and two, how to correctly estimate the generalization error for optimal parameters.

We can begin with the first issue, which essentially corresponds to choosing which parameters we want to vary, which range we'd like to consider, and which values we want to try out. This is really just a regular parameter optimization problem in which the parameters are correlated, so we have to vary all of them at the same time, exploring a multidimensional parameter space. It is easy to see the effect of the curse of dimensionality, especially because for each parameter combination, we have to run the model several times to obtain a reliable estimate of the performance through a cross-validated strategy. For these reasons, it is really important to *think* about the problem we are attempting to solve: If we are trying to decrease model complexity, we can limit our parameter space by excluding models more complex than the one under consideration; if multiple parameters have similar meaning or effect, we can vary only one of them.

### 4.5.1 Grid search

The most common approach to selecting hyperparameters consists of creating a grid of possible hyperparameter combinations, training a model in a CV framework for each of them, and *selecting the combination that yields the best test scores.*

For our SVM, we can decide to try out, besides the linear one, the polynomial and Gaussian kernels, and to co-vary parameters $d$ (the degree of the polynomial), $\gamma$, and C. From an implementation perspective, this can be achieved for example by building a dictionary of parameters, and then using the `GridSearchCV` utility from `sklearn.model_selection`, which runs a grid search within a cross-validation (CV) framework.

How should we choose the parameter values? They vary so much from algorithm to algorithm that it's hard to give a general rule, but practical advice borrowed from parameter estimation still applies:

- I would suggest starting with wide coverage (e.g., in $\gamma$ and $C$ we might want to span several orders of magnitude) and then, if necessary, refining the spacing of the grid;
- If the best parameters are found at the edge of the grid, we might want to further expand the explored range of parameters;
- If the grid is too large, `sklearn` offers the option of using a random search through `RandomSearchCV` instead;
- In principle, one could use any efficient parameter search technique, for example, Markov Chain Monte Carlo sampling, to find the best combination. However, in my experience at least, excessive fine tuning of the parameters is almost always rewarded with poor generalization properties (high variance), so a coarse-grained grid is in most cases just fine.

For our problem, we can use a pipeline to make sure that data are scaled before optimizing the parameters:

```
piped_model = make_pipeline(StandardScaler(), SVC())
```

and then create the following dictionary of parameters to explore:

```
parameters = 'svc__kernel':['poly', 'rbf'],
'svc__gamma':[0.00001,'scale', 0.01, 0.1], 'svc__C':[0.1, 1.0, 10.0, 100.0],
'svc__degree': [2, 4, 8]
```

This is just an example, and you might want to experiment with different ranges or reference values. Additionally, you might notice that I did not include the linear kernel in the list of kernels I used. There is a technical reason for this; `sklearn`'s User Guide recommends using the `LinearSVC` class for speed rather than `SVC` with a linear kernel, but most importantly, we can easily mimic the linear kernel behavior by either using a tiny value of $\gamma$ in the Gaussian (`rbf`) kernel (as we did above) or including degree $= 1$ in the polynomial kernel. These strategies are more efficient.

Finally, we can run the search as follows:

```
model = GridSearchCV(piped_model, parameters,
              cv = StratifiedKFold(n_splits=5, shuffle=True),
                return_train_score=True)
```

The command above results in exploring 120 parameter combinations, and because each one of them is used in a 5-fold CV scheme, the algorithm is trained 600 times. It is easy to see why choosing parameters carefully is essential to avoid the curse of dimensionality. Using the option `return_train_score=True` will help diagnose high bias/high variance issues in the optimized model.

If hyperparameter optimization is simply prohibitive to run on the full data set, we have the option of using a smaller random selection of the data to keep run times manageable. However, we have no guarantee that the optimal parameters will not change with data set size; this is particularly important if the learning curves have not flattened, so this strategy must be used with caution.

### 4.5.2 Analysis of results

Once our grid search run is complete, we can analyze the results. We certainly want to look at the winning model and its scores, but it's also very important to develop a sense of how different parameters are affecting the scores and whether the issues (variance/bias) of the algorithm have changed from our benchmarked case.

For example, one can build a data frame with the results of the grid search (contained in the attribute `model.cv_results_`) and look at the first 10 or 20 models; I usually like to look at the mean/standard deviation of test scores, at train scores, and sometimes at the time required to fit a model if it is relevant (if the data set is large or the algorithm is slow).

In this case, the scores look like those shown in Figure 4.11 (assuming that you used our same random seed in the CV; otherwise, you might get slightly different results), and they provide several insights:

1. All top 10 models use the Gaussian kernel and have very similar test scores;
2. The winning model's scores show only a slight improvement over the benchmark linear model, and the difference is not statistically significant;
3. The winning model(s) suffer from moderate high variance (the gap between train/test scores is significant at about the $3\sigma$ level).

Also note how the scores for the `rbf` kernel are identical for blocks of three models, because the three possible values of the "degree" parameter correspond to the same model.

Note that observation 1 doesn't mean that the overall performance of the algorithm is insensitive to the choice of parameters, but only that our benchmark model was already quite close to optimal. Scrolling down the data frame (shown in the notebook "ParticleID_wSVMs.ipynb"), we can see that other combinations of parameters (e.g., all the models with a polynomial kernel and $\gamma = 10^{-5}$) perform much worse and effectively correspond to the "lazy" classifier that places everything in the negative class. Verifying that models behave as expected as a function of varying parameters is a good sanity check that shows we didn't make a mistake in writing our code.

| | params | mean_test_score | std_test_score | mean_train_score |
|---|---|---|---|---|
| **43** | {'svc__C': 1.0, 'svc__degree': 8, 'svc__gamma': 'scale', 'svc__kernel': 'rbf'} | 0.8948 | 0.007626 | 0.92150 |
| **27** | {'svc__C': 1.0, 'svc__degree': 2, 'svc__gamma': 'scale', 'svc__kernel': 'rbf'} | 0.8948 | 0.007626 | 0.92150 |
| **35** | {'svc__C': 1.0, 'svc__degree': 4, 'svc__gamma': 'scale', 'svc__kernel': 'rbf'} | 0.8948 | 0.007626 | 0.92150 |
| **37** | {'svc__C': 1.0, 'svc__degree': 4, 'svc__gamma': 0.01, 'svc__kernel': 'rbf'} | 0.8944 | 0.008237 | 0.90015 |
| **29** | {'svc__C': 1.0, 'svc__degree': 2, 'svc__gamma': 0.01, 'svc__kernel': 'rbf'} | 0.8944 | 0.008237 | 0.90015 |
| **45** | {'svc__C': 1.0, 'svc__degree': 8, 'svc__gamma': 0.01, 'svc__kernel': 'rbf'} | 0.8944 | 0.008237 | 0.90015 |
| **61** | {'svc__C': 10.0, 'svc__degree': 4, 'svc__gamma': 0.01, 'svc__kernel': 'rbf'} | 0.8930 | 0.004382 | 0.91265 |
| **53** | {'svc__C': 10.0, 'svc__degree': 2, 'svc__gamma': 0.01, 'svc__kernel': 'rbf'} | 0.8930 | 0.004382 | 0.91265 |
| **69** | {'svc__C': 10.0, 'svc__degree': 8, 'svc__gamma': 0.01, 'svc__kernel': 'rbf'} | 0.8930 | 0.004382 | 0.91265 |
| **105** | {'svc__C': 1000, 'svc__degree': 4, 'svc__gamma': 1e-05, 'svc__kernel': 'rbf'} | 0.8912 | 0.007626 | 0.89460 |

Figure 4.11: **The first ten models, ordered by decreasing test scores, obtained by the grid search CV. Each row shows the hyperparameters of the model, the test scores and the standard deviation of the scores obtained from the multiple folds of the CV process, and the train score. This information helps us find out which parameters are the most important and whether our models suffer from high bias or high variance (from the gap between train and test scores, accompanied by the standard deviation of test score that helps us assess whether it is significant). Note that the top three models are identical (they only differ for the degree of the polynomial, but a Gaussian kernel is used; the grid search is not smart enough to recognize moot parameters).**

### 4.5.3 Nested cross validation

The grid search (or equivalent) procedure described above takes care of aspect one: picking optimal parameters. The second issue, as mentioned earlier, is how to correctly estimate the performance of our algorithm (classifier, in this case, but the scheme applies to regression problems as well). Should we just trust the test scores above? After all, as usual, we did our CV procedure and didn't peek at the test labels.

The problem in this case is that we have used the test scores as a litmus test to *choose* our optimal parameters. In other words, the optimization procedure was not blind to the test labels; the dreaded leakage of information has happened! Thus the test scores obtained in the grid search may be an overly optimistic assessment of what would happen on new data, where we don't have the luxury of retuning parameters, but we have to go with our best estimate based on the learning set.

The correct procedure to estimate the generalization error (the test scores on unseen data) is to use two *nested* levels of $k$-fold CV. The outer folds are used to split the learning set into train/test $k$ times, as usual. Then, *inside each of the* $k-1$ *outer training partitions*, we can repeat an $n$-fold CV scheme. The outer train set becomes the inner learning set, which is split into inner train/test folds. The grid search (or equivalent) hyperparameter optimization scheme is run within this inner learning set, and the performance *on the outer test fold*, which is now truly unseen, is the proxy for the generalization error. By repeating the process $k$ times, we obtain $k$ "winning models"; note that they might not always be the same, which is fine, because this procedure is only meant to estimate generalization error, not to pick the optimal model. The average and standard deviation of the $k$ winning models provide the estimate of how an optimal model would perform on unseen data.

This process is illustrated in Figure 4.12; in this case we have 5 outer folds (blue = test / gray = train), and in each of the 5 inner training sets, we run a hyperparameter optimization scheme with 4-fold CV (green = test / gray = train). The average (or median) of the scores on the 5 blue (outer test) folds is the generalization error. Code implementation is not currently provided in the lecture notebook.

Unfortunately, this procedure is extremely expensive in terms of computational resources; for example, testing 120 models (as we did above) in a nested CV scheme with 5 outer folds and 4 inner folds (the very minimum!) requires training the system *1,200 times*. For large data sets or more complex algorithms, this can quickly become prohibitive, making it even more essential to select wisely the parameters to tune and their range and to consider using a smaller sample selection where possible.

### 4.5.4 Picking the best model

So if we were to stop now in our optimization process, what would be the best model to pick? One would be tempted to just go with the one on the first line of Figure 4.11, but it probably wouldn't be the best choice. As already noted, the formal

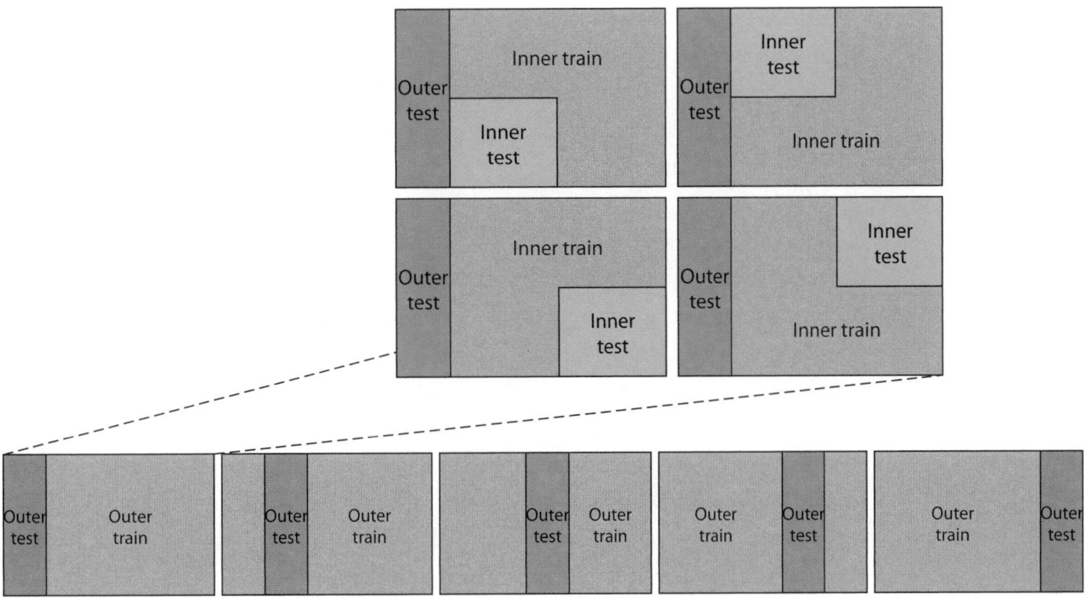


**Figure 4.12: A graphical example of the nested CV procedure. The outer CV schemes divide the learning set into 5 folds (blue = test, gray = train). Each of the gray training sets becomes the learning set for the inner 4-fold CV, in which hyperparameters are optimized through grid search. The scores on the 5 blue (outer test) folds, which did not participate in the hyperparameter selection process, provide an estimate of the generalization error and its uncertainty.**

winning model suffers from (moderate) high variance. However, our benchmark linear model (approximately represented in the grid by the combination $\gamma = 10^{-5}$, $C = 10^3$) has comparatively good test scores and does not suffer from high variance. We have repeatedly noted that the test scores are what matter, but when they are statistically equivalent, Occam's razor teaches us that a simpler model is always to be preferred. Additionally, the linear representation of the SVC algorithm has better time scaling than the other kernels. Therefore, at this stage, I would say that a linear model would be the winner.

## 4.6 FEATURE ENGINEERING

The procedure followed has told us that our "best" classifier so far suffers from high bias. Through our parameter optimization procedure, we have learned that increasing model complexity within the same algorithm (e.g., through Gaussian or polynomial kernels) is not an efficient solution, because it doesn't significantly improve the test scores while introducing a high variance issue.

We can try another route to improve, by engineering different features that hopefully will help "present" the information to our classifier in a better manner.

There are several flavors of feature engineering. The first one, and in my opinion the most powerful one when available, is to build features that "make sense" from a physical perspective. These are usually nontrivial combinations of available features that we might deem to be significantly correlated with the output; we will see some examples later in this section.

There are also "automated" methods to engineer features. For example, one can start from current features and build polynomial combinations up to certain degrees, ratios, differences, and so on. With kernel methods like SVMs, the process of feature engineering of this type is somewhat built into the kernel trick, because choosing a certain kernel function corresponds to choosing an enhanced set of features (e.g., all polynomial combinations up to a certain degree for the polynomial kernel, as discussed in Section 4.2.3). However, other methods, for example, decision trees can only operate on the basis of single features (all splits are parallel to existing axes). For such methods, engineering complex feature combinations is more likely to have a beneficial effect. Featuretools[3] is an interesting framework to automatically generate "intelligent" features in Python.

For our problem, we can use some knowledge of particle physics to develop additional features. We will consider some of them here, based on simple considerations about the possible decay channels described in Section 4.1.

The first variable we engineer is the total number of particles detected in the event. We would expect that 4-top events lead to a higher number of products, from the decays of four particles instead of two.

We can also use the number of product particles grouped *by type*. In particular, 4-top events are expected to generate on average a higher number of leptons (electrons, positrons, and negative and positive muons).

Similarly, we might expect to find a difference in the number of product particles in terms of photons, b-jets, and jets. So we will add the count for each particle type to the list of variables.

Overall, our simple analysis yields five new features: the total number of particles, leptons, photons, jets, and b-jets produced in each event. Many other possible combinations (e.g., taking into account the energy of the produced particles per type) could be added to this list.

With our updated list of features, we can repeat the procedure of benchmarking a reference model and then carry on with parameter optimization.

With the same linear model, standard scaling of features, and no regularization, we obtain a new test score of $0.948 \pm 0.007$. This result is far superior to the score we had obtained before adding the new variables; remember that the baseline accuracy shouldn't be 0 but rather the result of a random classifier that only knows about class size (around 73%), so this improvement is rather significant. Looking at the

3 https://www.featuretools.com/

train scores returned by the function `cross_validate`, note that the train and test scores are basically identical and so this model doesn't suffer from high variance issues.

We can reoptimize the hyperparameters of the new model (with updated features) by running the same grid-based search as described earlier in the chapter. The results are very similar; in this case as well, the reference model's performance is nearly optimal, and we might want to pick the best-performing linear model instead of the "official" winner (a Gaussian kernel model, in this case), because they have similar performances, but the former is simpler and faster.

### 4.6.1 Encoding categorical variables

Now that we are energized by the score boost provided by our savvy feature engineering process, the next question is: Can adding more features keep improving the scores? There is definitely information in our data set that we haven't harvested yet. For example, as we have learned that the types of particles produced in each event are important, we might want to add the variable type for each of the four product particles we are considering. This approach comes with an additional hurdle to clear: How do we add these categorical variables to an array that can be "fed" to the `sklearn` algorithms?

This is a common issue that can be solved in different ways. The simplest one is to just turn each categorical variable into a number, like we did for the target vector using the `LabelEncoder` utility. However, when used for more than one category, this method creates an implicit "distance" between numerical values: if a b-jet is categorized as a 0, a photon as a 1, and an electron as a 7, algorithms might interpret this as b-jets and photons being more similar to each other than to electrons, and events that produce b-jets as more similar to events that produce photons than to those producing electrons. This behavior might or might not be useful. In some cases, this similarity scheme is desirable, in particular when classes have some notion af adjacency that we might want to retain (e.g., days of the week or months of the year, although all these have the problem of cyclicity). If this is the case, one can specify the mapping of each category to a numerical value, for example, using `OrdinalEncoder` in `sklearn.preprocessing`, rather than relying on `LabelEncoder`, which assigns a 0 to the first variable type it encounters, 1 to the second one, and so on.

In our case, we'd rather not introduce a notion of similarity through the numerical encoding of particle types, so we can make use of the so-called *one-hot encoding* technique. The idea is that for each categorical feature we want to add, we add as many columns (features) as possible values of that feature. For example, the particle type of the first product ("Type_1") can assume the values "b" or "j". To one-hot encode this variable, we can add two columns, let's say "Type_1_b" and

"Type_1_j," and place a 0 or 1 in each of them to indicate whether the particle is a b- or a j-jet. The second product particle type can assume eight possible values (0/b/e+/e−/g/j/m+/m−), so its one-hot encoding will result in eight additional columns; their numerical values will be 0 for all of them except the one that matches the particle type. The process continues for all the particle types we'd like to encode (in this case, we can do it for the entire data set and then select which particle types we'd like to include in our feature list). There are several ways to achieve this with `python` or `sklearn`, but a really convenient one is by using `pandas`'s function `get_dummies`:

```
features_add = pd.get_dummies(data=features,
            columns=[`Type_'+str(i) for i in range(1,14)])
```

If we focus on the four particle products we are using, this process adds a total of 26 columns. However, the usual benchmarking and grid search process reveal that no further improvement comes from the inclusion of these new features. The best-performing linear model still presents no high-variance issue.

## 4.7 FURTHER DIAGNOSTICS AND FINAL MODEL SELECTION

Finally, we can ask whether including the additional information about all the other product particles and their types might help raise our test scores. As mentioned in Section 4.3.1, the problem here is that as we include the properties of additional particles (up to 13) in our feature list, we have a more and more difficult imputing problem. Imputing a constant value is not an appropriate solution when there are many missing values, because it introduces a bias and artificially reduces the variability of each feature.

In fact, let us see what happens when we use the most unsophisticated imputing technique, using a "0" value for all missing values. (You are invited to consider better imputing strategies in the exercises at the end of this chapter.)

If we add all other features (including the encoding of categorical "Type" variables), we end up with 156 variables. Using the same "benchmark" linear model as in the previous sections, we obtain test accuracy scores of $0.94 \pm 0.01$, in line with our previous results. No additional improvement comes from adding this information, and running our usual grid search for hyperparameter optimization reveals that the linear model is, in fact, optimal for these variables. A more interesting observation is that because of the additional noisy/uninformative features, *all models now suffer from high variance*, in different amounts, at the level of 1.5 standard deviations for the linear or quasi-linear ones, and much higher for Gaussian or polynomial kernel models with low regularization (high C or $\gamma$ values). This is unsurprising, and it means that in their current form, the new variables we have added are detrimental to the algorithm, because they only contribute to adding variance and increasing computational time without producing an improvement of the test scores.

Therefore, the best model among those considered would be the 18-feature linear model of Section 4.6, because it ties with more complicated models for the highest test scores (within the uncertainty) while having the lowest amount of variance.

It is understood that further preprocessing (e.g., using smarter imputing techniques) might change this assessment.

## 4.8 LESSONS LEARNED

Well this was a looooong chapter! I will divide our summary into the algorithm-level part and the more general methodology part.

Algorithm-wise, here is what we have learned:

- We now know all about Support Vector Machines (SVMs), our first kernel method.
- Kernel methods are awesome, because they have built-in feature engineering that amplifies the feature space but only require evaluation of the kernel function.
- SVMs offer a lot of flexibility but remain efficient, because the decision is determined by (often a limited number of) support vectors, not the entire data set.
- SVMs are a heavily parametric method for which scaling matters.

If you are interested in the mathematics behind the SVM algorithm you may want to look at the formal derivation of the solution to the optimization problem (Eq. 4.7). In a nutshell, the primal problem of finding the ideal coefficient in the same equation can be replaced by its *dual* formulation in terms of an appropriate Lagrangian, which can be solved through the sequential minimization algorithm. I recommend exploring references [Bennett and Bredensteiner, 2010; Schölkopf and Smola, 2001; Ben-Hur and Weston, 2010; Ng, 2019], which also address kernel theory and the necessary conditions to find an appropriate kernel function.

Methodology-wise, we can also summarize some relevant points:

- We applied our full problem-solving pipeline, from exploration to preprocessing to optimization, to a problem.
- We saw an example of a high-bias model and learned how content-based feature engineering can be useful for alleviating this issue; we also saw how noisy features may add variance to a model.
- We found that imputing matters and have discussed different strategies that will be implemented in future chapters.

- We discussed a few ways of encoding categorical variables as numbers, and the difference between turning variables into numbers (label encoding), which creates an implicit notion of distance, and adding columns with "checkboxes" (one-hot encoding), which avoids this issue.

In the next few chapters, we will learn how to set up and evaluate regression problems, where the target property is a continuous variable, and we will continue our exploration of different algorithms, starting from linear models.

## 4.9 REVIEW AND DISCUSSION QUESTIONS

Note: Questions and exercises marked by ** are more complex, open-ended, or time consuming.

**Exercise 1.** The separation between classes in an SVM is called

A) Boundary
B) Kernel
C) Support vector
D) Gamma

**Exercise 2.** Which feature of an SVM is closely related to the mapping between the original feature space and the enhanced, mapped feature space?

A) Gamma
B) C
C) The type of kernel
D) The class weight

**Exercise 3.** A large regularization parameter $C$ in an SVM means that . . .

A) We are giving misclassifications near the boundary a small penalty but gain a larger separation between classes.
B) We are giving misclassifications near the boundary a large penalty but will have a smaller separation between classes.
C) We are using a kernel with many degrees of freedom.
D) We expect a large generalization error.

**Exercise 4.** You are dealing with a binary classification problem in an imbalanced data set. You decide to use an SVM to solve it, but have limited time and can't afford to run a grid search for all the parameters, so you can only optimize one of them. Which one would you pick?

A) The regularization parameter $C$
B) The type of kernel
C) The degree of the polynomial kernel
D) The class weight

**Exercise 5.** Which of these kernel functions for an SVM has the highest risk of overfitting?

A) Linear SVM
B) Degree-2 polynomial SVM
C) Degree-5 polynomial SVM

**Exercise 6.** Which of these kernels of an SVM has the highest risk of high bias?

A) Linear SVM
B) Gaussian kernel SVM with a large $\gamma$ parameter
C) Degree-3 polynomial SVM

**Exercise 7.** To optimize the parameters of an algorithm, you should (write T/F):

A) Optimize all the parameters simultaneously.

B) Make sure there is no information leakage between the parameter optimization process and the test scores.
C) Optimize the parameters one at a time.
D) Give them all a pep talk.

**Exercise 8.** Nested CV is used to . . .

A) Estimate the scores achieved by optimizing model parameters.
B) Decide between classification and regression.
C) Build more training samples.
D) Turn unsupervised into supervised learning.

**Exercise 9.** You are doing nested CV and you have an outer loop with 5 folds and an inner loop with 3 folds. After selecting parameters in the inner CV loop, you report as generalization error. . .

A) The best score obtained in the 15 folds.
B) The average score obtained in the 5 outer test folds, corresponding to the model optimized on the inner 3 folds.
C) The average of all the scores in the 15 folds.
D) The worst score, just to be on the safe side.

## 4.10 PROGRAMMING EXERCISES

**Exercise 1.** For each kernel type, use the scores obtained in the grid search to discuss the impact of various parameters.

**Exercise 2.** Apply and optimize an SVM model to the planet habitability data set of Chapter 3. Make a wager first on whether the SVM would make for a better or worse model, compared to DT and kNN.

**Exercise 3.** **Experiment with different imputing strategies in this chapter's data set to harvest the information in the sparse features.

**Exercise 4.** **A few years ago, Kaggle, a well-known ML platform, ran a competition on learning how to "spot" the Higgs boson in simulated LHC data. Build a classification model on this tricky data set, following the prompts in the "HiggsBosonEx.ipynb" notebook.

# 5 CHAPTER FIVE
# Regression

*Shall I compare thee to gradient descent?*
*Probably not, that would be reaching a new low.*

In this chapter, we turn our attention to regression problems, the branch of supervised learning where the target property is a continuous variable.

We will start our exploration by examining linear models, perhaps the simplest class of regression models. We will discuss how the evaluation of these models proceeds by means of a *cost* or *loss function,* and we'll learn how the cost function is minimized during the training process. We will introduce the workhorse of optimization algorithms, a technique known as *gradient descent,* and discuss several variations. This analysis will also be useful when we discuss more complex models, such as neural networks, in Chapter 8. We will derive some practical considerations for improving regression models by means of the bias-variance decomposition and learn how regularization techniques can increase their robustness to generalization.

Finally, we will consider a more inclusive class of linear models, known as *generalized linear models,* which include well-known algorithms such as logistic regression and Poisson regression.

## 5.1 FROM CLASSIFICATION TO REGRESSION: WHAT'S NEW IN THE ANALYSIS PIPELINE?

At the end of Chapter 3, we established a tentative problem-solving pipeline for supervised ML models, whose main steps were preprocessing, dimensionality reduction, choice of an evaluation metric, benchmarking, diagnosing, optimizing hyperparameters, further diagnostics and adjustments, and final model selection, also shown in Figure 3.8. What changes would we need to make to tackle a regression problem? Luckily for us, very few: the main difference would be that we need to define different *evaluation metrics.* In a regression problem, our goal is to predict a value as close as possible to the true one, so we need a measure of *distance*, while in classification problems, we simply counted the number of correct or incorrect answers in each class. If our target value is 1.5 and our model predicts 1.498, we certainly won't call it wrong!

### 5.1.1 Evaluating regression models

Most metrics used for regression are based on the concept of residuals, the differences between the vector of predictions and the vector of true values. A perfect algorithm will exhibit a null residuals vector. In real-life cases, we can encode the performance of the algorithm using the mean (or median) absolute error (MAE), or the mean square error (MSE), or perhaps its square root, the root mean square error or RMSE. For an unbiased estimator (one where the mean of the residuals vector is zero), the MSE and RMSE correspond to the variance and standard deviation of the residuals vector, respectively.

If $y$ is the vector of true values, and $\hat{y}$ is the vector of predicted values, of length $N$, these metrics read:

$$\text{MAE} = \frac{\sum_{i=1}^{N} |(y_i - \hat{y}_i)|}{N}; \quad \text{MSE} = \frac{\sum_{i=1}^{N} (y_i - \hat{y}_i)^2}{N};$$

$$\text{RMSE} = \sqrt{\frac{\sum_{i=1}^{N} (y_i - \hat{y}_i)^2}{N}}. \tag{5.1}$$

Metrics of the type "the smaller, the better," which are based on the differences between predictions and true values, such as the MSE and the MAE, are called *loss functions.*

Another very common estimator for the quality of a predictive model is the $R^2$ score, or *coefficient of determination,* which describes the fraction of variability (variance) in the dependent variable that is explained by our predictive model. It can be expressed as a function of two variables, the *sum of total squares,* which is proportional to the variance of $y$ and quantifies the intrinsic variability in the data, and the *sum of squares of residuals,* which measures how well the predictions fit the data:

$$\text{SS}_{\text{tot}} = \sum_{i=1}^{N} (y_i - \bar{y})^2; \qquad \text{SS}_{\text{res}} = \sum_{i=1}^{N} (y_i - \hat{y}_i)^2, \tag{5.2}$$

where $\bar{y}$ is the mean value of the true values $(y)$ vector. The coefficient of determination is simply

$$R^2 = 1 - \text{SS}_{\text{res}}/\text{SS}_{\text{tot}}. \tag{5.3}$$

In a perfect model, each prediction $\hat{y}_i$ coincides with the true value $y_i$; the residuals vector is null, $SS_{res}$ is zero, and $R^2 = 1$. This is the best case scenario. A model that predicts the mean of the observed target values every time has an $R^2$ score of zero: the model can't explain any of the variance in the predictions. However, despite its name, the $R^2$ score can actually be negative, because the predictions of a model on the test set can be arbitrarily bad (worse than always predicting the mean!). The $R^2$ score for a linear model is only bound to be nonnegative *on the training set.*

In most prepackaged ML libraries, including `sklearn`, the $R^2$ score is the default evaluation metric for regression algorithms. However, it is worth noting that unlike in classification problems, there is no straightforward quantitative interpretation of the $R^2$ score.

For example, an $R^2$ score of 0.9 does not give any quantitative assessment of how close the predicted values are to the true values, and in particular, it does not mean that the predicted answer is typically within 10% of the correct value. This lack of an "absolute" interpretation is a common woe of all measures of correlation. Additionally, it is always possible to increase the $R^2$ score on the training set by adding new features, as long as they are not linearly dependent on one another, and by transforming the dependent variable (e.g., taking its log), we can easily change the $R^2$ score without changing the model at all. For these reasons, some notable statisticians are vocal critics of the $R^2$ score (e.g., see these notes[1]).

In my opinion, the $R^2$ score is a useful tool, despite its limitations. I find it helpful when *comparing* different models that have identical feature/target setups, in feature selection schemes, and in the process of optimizing a regression model. My takeaway message from the debate around $R^2$ is that evaluating regression models is a nontrivial task, and that monitoring several metrics at once may give us a better understanding of the strengths and weaknesses of our models.

In the following sections, we will discuss several additional assessment metrics.

## 5.2 LINEAR REGRESSION

Linear regression assumes the existence of a linear relationship between a dependent variable $y$ and some independent variables, often called *predictors*, $X$. The predictors correspond to the features in ML jargon, while the dependent variable is the target. If we have more than one predictor, we talk about *multilinear regression*.

In the context of supervised learning, our learning sample contains some observations of the independent variables, $X_{ij}$, while the $y_i$ are observed (sampled) values of the dependent random variable $y$. The linear model describes the response

1 http://www.stat.cmu.edu/~cshalizi/mreg/15/lectures/10/lecture-10.pdf

variable as a linear function of the predictors $X_{ij}$:

$$y_i = \beta_0 + \sum_{j=1}^{k} \beta_j X_{ij} + \epsilon_i, \tag{5.4}$$

where the $\beta_j$ variables are scalars called coefficients or *weights*, and $\epsilon$ accounts for the intrinsic noise of the observations.

### 5.2.1 Ordinary least squares

We have already spoken about "fitting a line" in Chapter 2 by minimizing the sum of squared deviations of the predicted values $\hat{y}$ from the observed values $y$. In the "language" of machine learning, fitting a linear regression model using the ordinary least squares (OLS) method corresponds to minimizing the MSE loss.

Note that in the classic statistical approach to linear regression, the validity of this process relies on some assumptions. Besides the linearity of the relationship, one needs to assume that the noise variables are independent and normally distributed with zero mean and constant variance (i.e., the error is independent of $X_i$); this property is known as *homoscedasticity*. As a result, for a given $X$, the displacement of the observed values $y$ from the expected value $E(y)$ of the response variable is only dictated by the variance of the noise term (e.g., see [James et al., 2013] and these notes[2]). A final necessary assumption is that the learning sample is representative enough of the entire population to be able to obtain good estimates of the coefficients by replacing the population parameters with the *sample* parameters and calculating statistical properties on the samples, instead of on the full distribution.

If the assumptions of the linear model listed above are satisfied, we are afforded a probabilistic interpretation of the results; for example, we can do *inference* on the parameters of the linear model and quote confidence intervals associated with them. Additionally, from an information theory perspective, we have some guarantee of *optimality*: The Gauss-Markov theorem ensures that (1) the coefficients of the OLS solution provide an *unbiased* estimator, and (2) among the unbiased estimators, the OLS solution leads to the optimal (minimum) MSE. Note that this doesn't make it necessarily the lowest MSE estimator, because of the bias-variance trade-off described in Section 5.5.

However, from an ML perspective, building a linear regression model (i.e., finding the coefficients that minimize the MSE loss and evaluating the performance of such model through CV) is still a perfectly legitimate operation. I will say, though, that if we are truly convinced that our data satisfy the hypotheses of a linear model,

---

2 http://people.duke.edu/~rnau/testing.htm

we should shoot for a statistical learning approach if at all possible. I enjoyed reading Lectures 1, 3, and 4 of Cosma Shalizi's lecture notes[3] for a rigorous treatment of this topic.

### 5.2.2 Analytic solution, one-dimensional case

The goal of the regression process is to estimate the coefficients $\beta_0, \beta_i$ of Eq. 5.4. Our starting point will be the loss function; for example, if we have chosen the MSE, we will want to find the coefficients that minimize the MSE on the training set. I show the derivation of the optimal estimates for the coefficients here for the case $j=1$ (one-dimensional data set, or fitting a line).

The MSE of our model can be written as

$$\text{MSE} = 1/N \times \sum_{i=1}^{N} (\beta_0 + \beta_1 x_i - y_i)^2, \tag{5.5}$$

which is, other than the factor $1/N$, the sum of squared residuals; minimizing the MSE loss corresponds to using the *least squares method.*

Finding the line that minimizes the MSE is a fun exercise that can be solved analytically. We can calculate the loss function's partial derivatives with respect to the coefficients, and then set them to zero:

$$\begin{aligned} \frac{\partial \text{MSE}}{\partial \beta_0} &= 1/N \times \sum_{i=1}^{N} (2\beta_0 + 2\beta_1\, x_i - 2\, y_i); \\ \frac{\partial \text{MSE}}{\partial \beta_1} &= 1/N \times \sum_{i=1}^{N} (2\beta_0\, x_i + 2\beta_1\, x_i^2 - 2\, x_i\, y_i). \end{aligned} \tag{5.6}$$

Setting the first term to zero (and simplifying $1/N$ away), we obtain:

$$\beta_0 = \frac{\sum_{i=1}^{N} y_i - \beta_1 \sum_{i=1}^{N} x_i}{N} = \bar{y} - \beta_1 \bar{x}. \tag{5.7}$$

We can now substitute this expression in the second term of Eq. 5.6, to obtain

$$\frac{\partial \text{MSE}}{\partial \beta_1} = 1/N \times \sum_{i=1}^{N} (2\,(\bar{y} - \beta_1 \bar{x})\, x_i + 2\, \beta_1\, x_i^2 - 2\, x_i\, y_i). \tag{5.8}$$

Setting this quantity to zero, we can solve for $\beta_1$:

$$\beta_1 = \frac{\sum_{i=1}^{N} x_i (y_i - \bar{y})}{\sum_{i=1}^{N} x_i (x_i - \bar{x})}. \tag{5.9}$$

3 https://www.stat.cmu.edu/~cshalizi/mreg/15/lectures/

It is easy to show that the above expression is equivalent to the following:

$$\beta_1 = \frac{\sum_{i=1}^{N}(x_i - \bar{x})(y_i - \bar{y})}{\sum_{i=1}^{N}(x_i - \bar{x})^2}, \tag{5.10}$$

which might be nicer as it recognizes that the best-fitting coefficient $\beta_1$ can be written as the sample covariance, Cov(X,Y) = $1/N \sum_{i=1}^{N}(x_i - \bar{x})(y_i - \bar{y})$, divided by the sample variance, Var(X) = $1/N \sum_{i=1}^{N}(x_i - \bar{x})^2$. Therefore, we get a physical intuition that the best fitting line in a linear regression becomes steeper when the response variable $y$ varies rapidly (has high covariance) with the predictor $x$, and is pulled toward a flat line if the sample variance (i.e., the range of the predictor) is large.

### 5.2.3 Analytic solution, *m*-dimensional case

In the multidimensional case (with, say, $m$ features), the model takes the form $y = \beta_0 + \beta_1 x_1 + \beta_2 x_2 + \cdots + \beta_m x_m$. It is customary to add a "bias" term (unfortunately, this has nothing to do with the bias associated with ML algorithms that I have often mentioned) as an additional feature with value $x_0 = 1$ to all objects in the data set. This allows us to write the model in compact matrix form, as the intercept term becomes the parameter that multiplies the new bias feature $x_0$:

$$\hat{y} = \mathbf{x} \cdot \beta. \tag{5.11}$$

Our goal is to find the $(m+1)$-dimensional vector of parameters $\beta_0, \ldots \beta_m$ that minimizes the MSE loss.

If our training set is composed of $N$ objects, just as before, the MSE becomes

$$\text{MSE} = 1/N \times \sum_{i=1}^{N} (\mathbf{x}_i \cdot \beta - y_i)^2. \tag{5.12}$$

Again, by setting the gradient of the MSE with respect to the parameters equal to zero, we can obtain an analytical solution:

$$\hat{\beta} = (\mathbf{X}^{\mathbf{T}} \cdot \mathbf{X})^{-1} \cdot \mathbf{X} \cdot \mathbf{y}, \tag{5.13}$$

where $\mathbf{X}$ is the feature matrix, containing $N$ rows and $m$ columns. The equation above is known as the *normal equation.*

## 5.3 LINEAR MODELS AND LOSS FUNCTIONS

Linear models are simple, and they offer a great opportunity to visualize what is going on under the hood in most ML algorithms, giving us a chance to understand the building blocks of more complicated models, such as neural networks.

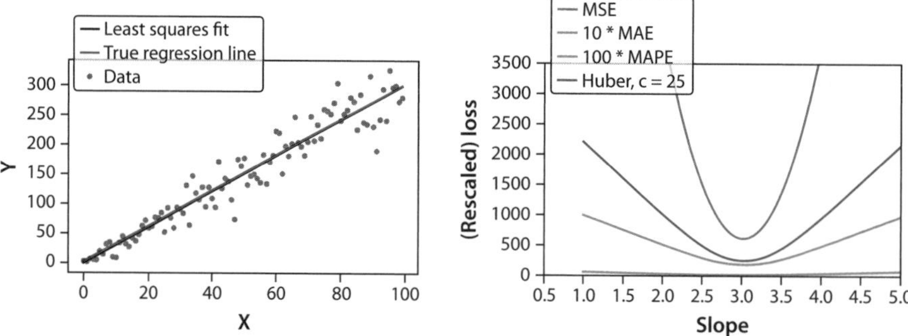


**Figure 5.1: (Left) Best-fit line for the MSE loss (OLS solution). (Right) Behavior of different losses as one moves away from the best-fitting solution. This figure also illustrates how different loss functions "weigh" outliers.**

In this section, we will set aside the predictive component of the model and focus on understanding how different loss functions are handled. You can follow along the text in this and the next section using the notebook "LinearRegression.ipynb."

To start, we can generate some two-dimensional data that follow a linear relationship ($y = 3x + 3$) and add some noise (note that we are purposely avoiding modeling the noise as Gaussian, to show that this doesn't matter if all we ask for is the line with the lowest possible MSE; things would be different if we wanted to claim that our model is correct and give it a probabilistic interpretation, i.e., claim that we have inferred the true parameter distributions). The data are shown in Figure 5.1.

To find the best regression line, a.k.a. the one with the lowest MSE, we could plug in the best coefficient we find using the formulas in Section 5.2.2 (in fact, you should calculate $\beta_0$ and $\beta_1$ from the formulas before continuing). However, just for fun, we can invoke `sklearn`'s linear model:

```
from sklearn import linear_model
model = linear_model.LinearRegression()
```

and the use the "fit" property of the model to fit our entire data set. The slope and intercept of the regression line are accessible as follows:

```
slope, intercept = model.coef_, model.intercept_
```

and they have the (rounded) values 3.025 and $-0.1255$, respectively. Hopefully, from your calculation using the formulas above, you found the same numbers (if you have used numpy's covariance and variance functions, you might have slightly different results; this is simply due to the $1/(n-1)$ factor used in numpy to de-bias the covariance estimator, which has a $\sim 1\%$ effect in our data set where $n = 100$). These numbers are not exactly the same as the ones we used to generate the data, but they are close enough (the intercept in particular is hard to get exactly right, because it makes a small contribution to the total MSE), so that the regression line found by the OLS method is quite close to the true regression line, as we can see in Figure 5.1.

### 5.3.1 Predictive linear models

After our sanity check on how to find the best regression line in a linear model, we can also, of course, apply the usual machinery to use the model in a predictive fashion through the CV process.

```
cv = KFold(n_splits=5, shuffle = True, random_state=10)
scores = cross_validate(model, x.reshape(-1,1), yp, cv = cv,
return_train_score = True)
```

The mean and standard deviations of the test "scores" reported are $0.91 \pm 0.04$ (with train scores $0.925 \pm 0.01$). But what do those numbers represent? Because we didn't specify a scoring parameter, `sklearn` reports as a default the $R^2$ score that we discussed earlier. An $R^2$ score of 0.9 and above sounds pretty good, but as mentioned, it doesn't have an immediate interpretability.

Therefore, we might ask ourselves what the MSE is of the best regression line, or even one of the other metrics, such as the MAE or median absolute percentage error (or MAPE, which doesn't get much love in statistics books, but I have found to be quite useful in scientific applications).

This part is easy enough: We can just change the scoring parameter in the `cross_validate` function, for example:

```
scores = cross_validate(model, x.reshape(-1,1), yp, cv = cv,
scoring = 'neg_mean_squared_error', return_train_score = True)
```

(note that the error-type scoring parameters in `sklearn` have a negative sign, so that they maintain semantic coherence with the idea of "scores," where higher is better).

Changing the scoring parameter helps us visualize the metric we have chosen, just as in classification problems, when we switched from the default accuracy to precision, recall, or a custom scorer. However, as noted earlier, *changing the scoring parameter does not change the way the best regression line is calculated.* The line stays the same in all cases, because the OLS minimization (which corresponds to minimizing the MSE) is "baked" into the linear model in `sklearn`. The only difference is which scores are *reported.*

So if we want to truly optimize the regression process with respect to a different metric (e.g., the MAE), what can we do?

One possibility would be to attempt to calculate its derivative with respect to the coefficients analytically, and set it to zero, as we did for the MSE. Sadly, a closed-form solution does not exist for this case. Therefore, we are better off looking for the minimum of the loss function using some other method.

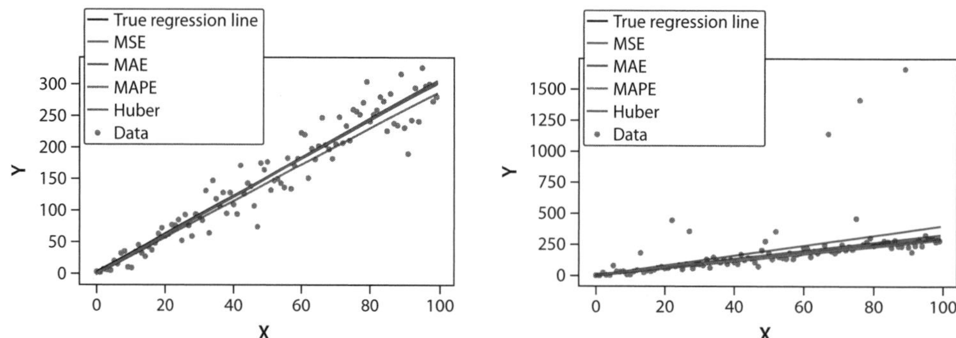


**Figure 5.2: The coefficients of the linear model vary as a function of the loss function that is minimized. The effect is larger when data include outliers; the MSE loss is more sensitive to outliers than the others, and its best-fitting line result is skewed away from the true regression line.**

The simplest one is very crude: We make a grid of possible coefficient values, calculate the loss function(s) of interest for all the combinations, and pick the values associated with the minimum. This is similar to what we did in Chapter 2 and also similar to the cross-validated grid search that we used when we learned how to tune hyperparameters in Chapter 4.

Let us repeat this exercise for different loss functions. Besides the MSE, MAE, and MAPE that we know already, results are also shown for the *Huber* loss, which is a hybrid between the MAE and MSE: Deviations from true values are penalized through a L2 norm (quadratic difference) up to a point controlled by the parameter $\epsilon$, and through a L1 norm (absolute difference) afterward. This achieves the goal of giving significant weight to incorrect measurements without being swayed too much by outliers. The different shapes of the loss functions as a function of the "slope" parameter of the grid search (keeping the intercept at fixed value, $\beta_0 = 0$, for visualization purposes) are shown in the right side of Figure 5.1.

The left panel of Figure 5.2 shows the regression lines obtained after finding the optimal parameters for these different loss functions. Compared to the original data, which exhibit very regular behavior, differences are minimal (on the order of a few percent). However, when data contain outliers, as shown in the right panel of the same figure, the choice of loss function becomes more important, and in particular, we can see how the MSE loss is quite sensitive to the presence of a handful of outliers.

## 5.4 GRADIENT DESCENT

For the simple two-dimensional problem we are considering, the grid search approach to minimization works really well; we are only looking at two parameters, so if our grid has 200 values/parameter, we can get away with 40,000 evaluations of the loss function, which is manageable. However, for more complex parameter spaces, we will need to employ smarter strategies to find the minimum of the loss function. The main idea behind such strategies is to explore the parameter space by moving *along the negative gradient of the loss function*, so that at each successive iteration, we hope to move toward a lower value. This method, which is at the core of most optimization strategies of ML algorithms, is called *gradient descent*. The examples you see here come from the notebook "LinearRegression_GradientDescent.ipynb."

### 5.4.1 Gradient descent for MSE loss

We can start by looking at gradient descent for a familiar case: MSE loss. The advantage here is that we will be able to compare the results to the analytic solution provided by the normal equation, Eq. 5.13.

The pseudocode for "standard" (often called *batch*) gradient descent is very simple:

- `Choose N_it = number of iterations and` $\eta$ `= learning rate;`
- `Select at random initial parameter values,` $\beta$`_t = 0;`
- `For n < N_it:`
  - `Calculate gradients of the loss function with respect to current parameter values:` $\nabla = \nabla_\beta L(\hat{y}(\beta))$`;`
  - `Update parameters by moving in the direction of the negative gradient by an amount determined by the learning rate:` $\beta = \beta - \eta \times \nabla$`;`
  - `Increment the number of iterations by 1.`

The learning rate is an important parameter of any gradient descent algorithm, because it controls the size of steps made in the space of variables. Steps that are too large will lead to instability, as the loss function will oscillate around its current value rather than move consistently toward the minimum; steps that are too small will slow down the convergence to the minimum of the loss function and require a larger number of iterations. As we will see in Chapter 8, using an adjustable learning rate (or more generally, an *optimizer*), is greatly beneficial. Still, for a convex loss function, gradient descent is guaranteed to converge if the number of iterations is sufficiently large.

The calculation of the gradients appearing in the gradient descent formulation above is a key element of optimization. Modern software packages rely on techniques such as *automatic differentiation* to automatically and efficiently calculate gradients. For the MSE loss, we already know from the previous sections that the gradients can be computed analytically. Therefore, batch gradient descent (BGD) can be written in a couple of lines of code as follows:

```
for n in range(N_it):
    grad = 2/m × X.T.dot(X.dot(β) - y)
    β = β − η × grad
```

Before comparing the results of BGD and the normal equation, we can ask: Why bother to use BGD for the MSE loss, given the availability of an analytic solution? The reason is that *the computational complexity of the normal equation is pretty brutal when the number of features is large*: It scales linearly with the number of examples in the training set, and polynomially, with an exponent between 2.5 and 3, with the number of features. In contrast, for the MSE loss, the computational complexity of BGD is linear with respect to the features and the number of examples in the data set. The main difference is that BGD requires only matrix multiplications, while the normal equation Eq. 5.13 requires inverting the feature matrix. As a result, *when the number of features in the data set is large, the BGD algorithm will be faster*. Additionally, as already mentioned, the BGD can be used for any loss, as the gradients can be computed numerically.

In the notebook, we apply the BGD algorithm with learning rate $\eta = 0.0001$ and 1,000 iterations to our data with outliers from Figure 5.2. The resulting values for the slope and intercept are 1.39 and 3.98, respectively; they are similar, but not identical, to the numbers obtained using the normal equation: 1.55 and 3.98. As noted earlier, the intercept is harder to pinpoint from linear regression, compared to the slope, because it has a smaller effect on the loss; in fact, the difference between the MSE losses corresponding to the two solutions is $\mathcal{O}(10^{-5})$. As expected, if we increase the number of iterations (e.g., to 100,000), we can recover the analytic result.

### 5.4.2 Stochastic Gradient Descent

Batch gradient descent is a faster alternative to the normal equation when the number of features is large, but it might still be computationally expensive when the number of samples in the data set is large. The bottleneck in this case is given by the large number of gradient evaluations, and the large matrix multiplication required to compute the loss at each iteration.

For large data sets, a popular alternative is given by *stochastic gradient descent* (SGD). The main idea is to use the gradient descent algorithm but compute the gradients *using only one instance at a time*, as opposed to all instances in the data set. For the MSE loss, the code is a slight modification of what we saw for BGD: For each iteration, the (X,y) pair for one random sample replaces the feature matrix/target vector:

```
for n in range(N_it):
    random_index = np.random.randint(m)
    X_one = X[random_index:random_index+1]
```

```
y_one = y[random_index:random_index+1]
grad = 2 × X_one.T.dot(X_one.dot(β) - y_one)
β=β-η × grad
```

As you can imagine, the path taken by SGD in parameter space is much less regular than the one taken by BGD, and in fact, while SGD is expected to generally move toward the minimum of the loss function, it is not guaranteed to find the minimum or to stay there after finding it. This apparent nuisance is the key to a greatly useful property of SGD, which is the ability to "jump" out of local minima of non-convex losses. The other claim to fame of SGD is, of course, its remarkable speed; a little complexity table, valid for strongly convex loss functions (informally, convex functions whose curvature does not approach zero), for BGD versus SGD is shown in Table 5.1.

### 5.4.3 Mini-batch gradient descent

Mini-batch gradient descent (MGD) is a compromise between batch and stochastic gradient descent, where at each epoch, we choose a certain number of examples, for example 10 or 20, at random, and use them to compute the gradients. The mini-batch size can be tuned as a hyperparameter. The advantages of MGD are

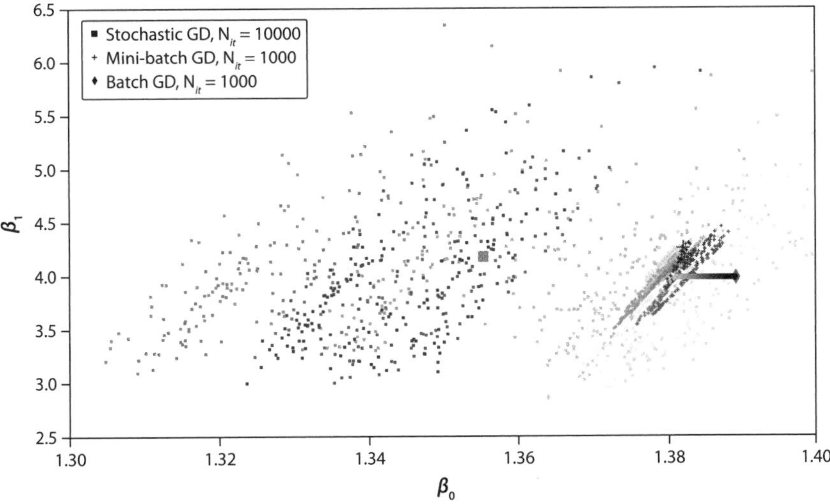


**Figure 5.3: Paths taken in slope/intercept space by batch, stochastic, and mini-batch gradient descent, for the same initial conditions. Color shades become increasingly darker as the number of iterations increases; the larger symbols show the final path. Note the larger scatter shown by SGD compared to MGD and the deterministic trajectory of BGD. Also note that the solution provided by the normal equation, with intercept $\sim$ 1.55, would not be shown in this plot; however, the difference in the minimum of the MSE loss function, compared to the solution of the normal equation, ranges from $\mathcal{O}(10^{-5})$ for BGD to $\sim$ 0.29% for SGD.**

Table 5.1: **Complexity table for BGD vs SGD, for a data set with $m$ features and $n$ instances, with a tolerance in the gradient norm of $\epsilon$. These scaling laws are valid for strongly convex loss functions. Adapted from [Zadeh, 2015].**

| Method | Number of iterations | Cost per iteration | Total cost |
|---|---|---|---|
| Batch Gradient Descent | $\mathcal{O}(\log(1/\epsilon))$ | $\mathcal{O}(nm)$ | $\mathcal{O}(\frac{n}{m}\log(1/\epsilon))$ |
| Stochastic Gradient Descent | $\mathcal{O}(1/\epsilon)$ | $\mathcal{O}(m)$ | $\mathcal{O}(m/\epsilon)$ |

pretty clear: On the one hand, we reduce the stochasticity from SGD, because the gradients computed on a selection of samples will track more closely the gradients computed on all samples, as for BGD. On the other, we still retain a better time complexity, compared to BGD, by only using a fraction of the data, and we are still able to navigate our way out of local minima. By tuning the size of the mini-batch, we move along the precision vs complexity axis, between the two limiting cases of SGD (mb_size = 1) to BGD (mb_size = $N$). Here is an example, where the size of the mini-batch is 10:

```
mb_size = 10
for n in range(N_it):
    shuffled_indices = np.random.permutation(m)
    X_shuffled = X[shuffled_indices]
    y_shuffled = yp_wo.reshape(-1,1)[shuffled_indices]
    x_mb = X_shuffled[:mb_size]
    y_mb = y_shuffled[:mb_size]
    grad = 2/mb_size × x_mb.T.dot(x_mb.dot(β) - y_mb)
    β=β−η × grad
```

The behavior of the three algorithms for the same optimization problem considered in the previous sections is shown in Figure 5.3. The path of BGD in the $(\beta_0, \beta_1)$ parameter space is very smooth and entirely deterministic. The path of SGD has much larger variance, and while it generally tends to move toward the correct solution, there is no guarantee of an improvement *at each step*. The path of MGD is intermediate between the two. A summary of the pros and cons of various implementations, inspired, like a lot of this chapter, by the excellent content in [Géron, 2019], is shown in Table 5.2.

In practice, the basic implementations shown here can be greatly improved, in terms of stability and performance, by setting a *learning schedule*, which is an adjustment of the learning rate that implements changes in step size as we progress toward the minimum in order to increase efficiency. All modern optimizers (e.g., Adam [Kingma and Ba, 2014]) include learning schedules that also depend on the current values of gradients and their trends; we'll revisit this topic in Chapter 8, when we talk about backpropagation for neural networks.

Table 5.2: **Some characteristics of linear regression methods for a data set with *n* instances and *m* features. Adapted from [Géron, 2019].**

| **Method** | **Large *n*** | **Large *m*** | **Hyper-params** | **Scaling?** | **`sklearn` implementation** |
|---|---|---|---|---|---|
| Normal Equation | slow | fast | — | no | `LinearRegressor` |
| Batch GD | fast | slow | yes | yes | — |
| Mini-Batch GD | fast | fast | yes | yes | — |
| Stochastic GD | fast | fast | yes | yes | `SGDRegressor` |

### 5.4.4 Gradient descent for generic loss functions

Even if we introduced gradient descent as an alternative to the normal equation for large data sets, its true power is in its flexibility. The gradient descent techniques mentioned above can be used for generic loss functions, including nonconvex ones. For such functions, the presence of a stochastic component and an intelligent learning schedule will be crucial for ensuring that we can move out of any local minima and converge toward the global minimum. The main difference in implementation is that for most loss functions, there is no closed-form expression for the gradient of the loss. In those cases, the gradients need to be computed numerically. Thankfully, evaluating derivatives of a function at a certain point in parameter space is a well-studied computational problem, which can be solved with automatic differentiation techniques. I liked [Baydin et al., 2018] for an introduction to the topic and a survey of how it is used in ML problems.

## 5.5 BIAS-VARIANCE TRADE-OFF

Previous chapters have mentioned that the process of optimizing an ML model often starts with establishing a "benchmark" performance and then diagnosing possible issues, such as high variance or high bias, which are associated with the concepts of overfitting and underfitting, respectively.

Now we are ready to make a more formal argument on how those indicators are connected to each other and explain why fixing a high bias or high variance issue can lead to an improvement in the overall generalization error, approximated as usual by the test error.

### 5.5.1 Bias-variance decomposition for MSE loss

The key idea is the so-called *bias-variance decomposition*, which is an easily-derivable result and states that the MSE loss *on test data* can be written as the sum of squared bias, variance, and noise [Geman et al., 1992; Hastie et al., 2001]:

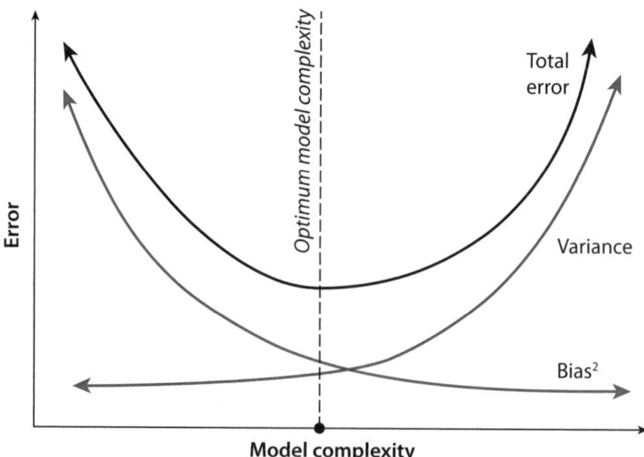


**Figure 5.4: Illustration of the "traditional" view of bias-variance trade-off for the MSE loss. Bias is expected to decrease with increasing model complexity, while variance is expected to increase. Because their contribution to the total loss is not equal and opposite, an "optimal" model complexity can be found. Figure from [Fortmann-Roe, 2012].**

$$\begin{aligned} \text{MSE (test)} &= b^2 + \sigma^2 + \mathcal{N}, \\ b &= E[y - \hat{y}], \\ \sigma^2 &= E[(\hat{y} - E[\hat{y}])^2], \end{aligned} \tag{5.14}$$

where the bias $b$ represents the average shift of the predictions, the variance $\sigma$ represents the range of variation of each prediction around the true value, and the noise $\mathcal{N}$ is an irreducible term that is model independent and determined only by the data.

The bias-variance decomposition is useful because it shows explicitly why exploring different combinations of bias and variance may lead to an improvement of the test score, but it doesn't, by itself, provide a strategy to lower the bias or the variance. This loftier goal is instead centered on the concept of *bias-variance trade-off*, which is the idea that *bias and variance for an algorithm are monotonic functions of model complexity*. Intuitively, the bias is expected to decrease as model complexity increases, as the model is able to accommodate more complex input/output relationships, and the variance is expected to increase, as the model acquires the capability to "absorb" and memorize many features of the training set, including spurious or noisy ones. This intuition is illustrated in Figure 5.4.

This empirical idea has been confirmed and embraced by many practitioners (see, e.g., [Geurts, 2009]) and allows one to forge a path for improvement, where high bias is alleviated by increasing model complexity, and high variance is alleviated by decreasing model complexity. This idea is relevant because *the trade-off is not exact*: A decrease of $x$ amount in squared bias does not imply an increase of the same amount in variance. Therefore, there is an expected "optimal" model complexity where the test MSE is minimal and our diagnostic tools help us move along the gradient of the total curve toward the optimal model.

Note that in the past few years, several practitioners have shown that for complex algorithms such as neural networks, the simple interpretation of bias-variance trade-off as a function of model complexity does not hold, and it depends additionally on whether the model is in an under- or over-parameterized regime [Belkin et al., 2019; Neal et al., 2018]; see also [Hogg and Villar, 2021]. They suggested that the curve in Figure 5.4 extends to a double-u shape when the overparameterized regime is included and showed that it is possible to obtain a decrease in variance without an increase in bias by moving to this region. This interesting topic is under active development in the ML community; let us set it aside for the moment, as our discussion about the role of regularization remains valid.

For linear models, for some reasonable conditions (if the errors in the linear regression model are uncorrelated, have equal variances, and expectation value of zero), the Gauss-Markov theorem tells us that the OLS solution is the optimal *unbiased* solution; in other words, the one that corresponds to the smallest MSE (and, since the bias is zero, to the lowest variance).

However, no conclusions can be drawn about *biased estimators*. The bias-variance decomposition indicates that we can potentially find some biased estimators that have lower total MSE, models in which a small increase in bias with respect to the OLS solution might be rewarded by a larger decrease in variance. The idea that biased estimators might lead to a lower MSE compared to the OLS solution makes it worthwhile to think about our next idea: *Regularization.*

Note that the bias/variance decomposition can be applied to other loss functions as well; for example, see [James, 2003] and references therein.

## 5.6 REGULARIZATION

Having a regression process that is robust to outliers is desirable for many reasons. One, we like our algorithms to be able to recognize objects as "unusual" and avoid putting excessive weight on them during the learning process. Two, in general, models that are not easily swayed by a handful of outliers tend to generalize better (in other words, sensitivity to outliers might correspond to a tendency to overfit). We have shown already how one can select different loss functions (which may or may not correspond to the evaluation metrics we use to compare different models) and perhaps pick a loss function that is more robust to outliers.

An alternative approach is to solve the OLS problem (i.e., use the MSE loss) and add one or more *regularization parameters*, whose purpose is to limit the size of coefficients of the regression process.

### 5.6.1 Flavors of regularization

We saw in Section 5.5 that the process of minimizing the MSE can be written as the minimization of the L2 norm of the difference between $\mathbf{X} \cdot \beta$ and $\mathbf{y}$:

$$\min_{\beta} ||\mathbf{X} \cdot \beta - \mathbf{y}||_2^2. \tag{5.15}$$

One of the most common regularization processes is the so-called *Ridge regression*, which solves the following modified optimization problem:

$$\min_{\beta} ||\mathbf{X} \cdot \beta - \mathbf{y}||_2^2 + \alpha \, ||\beta||_2^2. \tag{5.16}$$

Note that the intercept $(\beta_0)$ is excluded from the coefficient vector $\beta$ in the second term, to ensure that the model is invariant to translation.

The addition of the second term in the loss function above ensures that the coefficients don't become too large, which can happen in the presence of outliers, as we saw above, but also when the features are highly correlated, as you can see in the "Regularization.ipynb" notebook.

The parameter $\alpha$ is akin to the "C" regularization parameter of Support Vector Machines discussed in Chapter 4 and can be optimized through cross validation, just like any model hyperparameter.

A desirable property of the Ridge regression algorithm, when used with the MSE loss, is that an analytic solution is still available, in the form of a simple modification of the normal equation 5.13:

$$\hat{\beta}_{\text{Ridge}} = (\mathbf{X}^{\mathbf{T}} \cdot \mathbf{X} + \alpha I)^{-1} \cdot \mathbf{X} \cdot \mathbf{y}. \tag{5.17}$$

Qualitatively speaking, the Ridge regression solution increases the magnitude of the eigenvalues of the matrix $\mathbf{X}^{\mathbf{T}} \cdot \mathbf{X}$ (because the coefficients $\alpha$ are strictly positive). If some of them are very small, they are effectively pulled away from zero, so that when this matrix is inverted, the coefficients $\hat{\beta}_{\text{Ridge}}$ tend to be smaller, compared to the nonregularized solution.

A popular alternative to the Ridge regression is the *Lasso* regression, in which the regularization term is proportional to the L1 norm of the coefficients vector:

$$\min_{\beta} \frac{1}{2N} ||\mathbf{X} \cdot \beta - \mathbf{y}||_2^2 + \alpha \, ||\beta||_1. \tag{5.18}$$

This strategy applies a stronger regularization compared to Ridge regression. Intuitively, when a feature (j) is only weakly correlated with the output $y$, its associated coefficient $\beta_j$ is small $(< 1)$, and this regularization is stronger for equivalent values of $\alpha$. In fact, Lasso regression tends to give sparse results and produce models with fewer nonzero coefficients. This makes it helpful as a feature selection method, and useful also when features exhibit high degrees of correlation. Often it is employed as a second step to feature engineering; for example, in a high-bias problem, one can generate many possibly noisy features to attempt to find some useful ones, and then employ Lasso regression to eliminate some of them, hopefully ending up with an improved feature set.

One last mention-worthy regularization strategy is the hybrid *Elastic Net*, which employs a mixture of L1 and L2 norms of the coefficient vector, with the degree of mixing regulated by a hyperparameter.

We can compare the effects of adopting a linear model without regularization, the Ridge regression and the Lasso regression in multicollinear data. We begin by generating a 3-feature sample (xb) following a linear relationship and then create new polynomial features from the original ones:

```
from sklearn.preprocessing import PolynomialFeatures
poly = PolynomialFeatures(2, include_bias=False)
new_xb = poly.fit_transform(xb)
```

The new data set has 9 features, and of course (given how they were created) they are not all independent of one another. Fitting the data with the simple linear model returns the following set of coefficients:

$$\begin{aligned}\text{Coef} = [&1978.380, 19.983, -1202.206, 15.683, 0.1584, 8.8872,\\ &0.0016, 0.08977, -0.1356].\end{aligned}$$

This behavior is less than desirable, because many of the numbers are large; the response variable will change significantly as a result of a small change in the corresponding input features. This pattern is often associated with overfitting.

In contrast, for the Ridge regression model with $\alpha = 1.0$ and $\alpha = 1000.0$, we obtain

$$\begin{aligned}\text{Coef} = [&214.19, 2.1635, -114.18, 0.59258, 0.00599, 1.1995, 6e-05,\\ &0.01212, -0.02956]\end{aligned}$$

and

$$\begin{aligned}\text{Coef} = [&7.15368, 0.07226, 11.07351, -0.98855, -0.00999, 0.29255,\\ &-0.0001, 0.00296, -0.01639],\end{aligned}$$

respectively; the amount of regularization increases with increasing $\alpha$, and the coefficients become smaller.

The Lasso regression exhibits a similar behavior but tends to favor sparsity in the coefficients by setting some of them to zero; for example, for $\alpha = 1.0$ and $\alpha = 1000$, respectively, we obtain

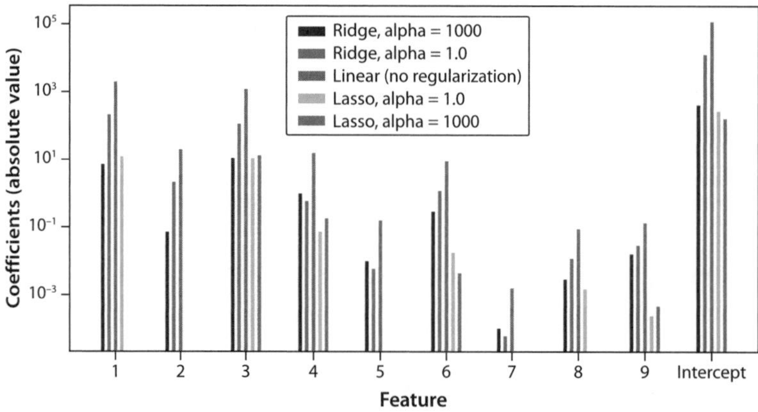


**Figure 5.5: The coefficients of the linear model vary significantly as a function of the amount and type of regularization applied. For the same value of $\alpha$, Lasso regularization is stronger and induces a sparse grid of coefficients, which makes it suitable for feature selection. Note that the y axis is in log scale and that the intercept does not participate in the regularization process.**

$$\text{Coef} = [12.42005, 0.0, 11.12834, 0.07434, 0.0, 0.01825, 0.0, \\ -0.00154, 0.00024]$$

and

$$\text{Coef} = [0.0, 0.0, 13.38545, 0.18356, 0.0, 0.00446, 0.0, 0.0, -0.00047].$$

The number of nonzero coefficients decreases as the strength of regularization increases.

It's perhaps easiest to visualize the effect of regularization by looking at a bar chart of coefficient values, shown in Figure 5.5. The original distribution of coefficients was $[3.0, 0.5, 15.0, 0, \ldots, 0]$; while none of the algorithms is able to reproduce them exactly, the regularized models (and in particular, Lasso) with significant regularization (large $\alpha$) are better at picking the important feature #3. Note that, as usual, the optimal value of $\alpha$ in the two cases should be decided by cross validation, and you are invited to do so in the exercises section.

It may also be interesting to test our intuition that models with stronger regularization will be more robust to generalization, as we move away from the model training domain.

## 5.7 GENERALIZED LINEAR MODELS

Generalized linear models (GLMs) aim to recycle the useful machinery of linear models while exploring more flexible relationships between the predictors and the response variable. Typically, this happens through a transformation (link function, e.g., a logarithm), and the transformed variable is assumed to be linearly dependent on the predictors. In the probabilistic formulation of GLMs, there are updated conditions that need to be met to afford a probabilistic interpretation of results: The response variable does not need to be described by a normal random variable, but it can be a member of any exponential family (Poisson, gamma, Bernoulli), and similarly, uncertainties do not need to be independent identically distributed but can also follow one of the above distributions. In our ML approach, we are less concerned with these assumptions, and we can think of GLMs as tools to model a wider range of input/output relationships within the useful framework of linear models, which includes optimization and regularization.

### 5.7.1 Logistic regression

One popular example of GLMs is the so-called *logistic regression*, which, confusingly enough, is in most cases used as a classification algorithm.

Logistic regression is used when the response variable has a limited range. In a regular linear model, if the values of linear predictors become very large or very small, the response variable will tend toward arbitrarily small or arbitrarily large values (in fact, $-\infty$ and $\infty$). However, there are many cases in which we might want to limit the range of the response variable; for example, if we are estimating the value of a used car on the basis of various features such as year of production and model, the response variable should not go negative, and similarly, there is probably a maximum resale value.

One particular case of limited-range outcome is when we are modeling *odds*: We are looking at an event with two possible, mutually exclusive outcomes (success/failure; happen/didn't happen, and so on), and we want to model the ratio of the odds of the two outcomes. For example, if I am looking at "raining" as a possible event, and I believe that there is a 75% probability that it will rain, then there is a 25% chance that it won't rain, and the odds are 3:1 in favor of the rainy outcome.

This problem can, of course, be solved as a classification problem. However, the extra advantage of the logistic regressor is that by modeling the probabilities of either outcome, we retain information about the degree of confidence in our answer. The odds of 1:1 (i.e., having a probability of 50% associated with each outcome) will act as the separator between the two classes, but we will feel differently about our prediction if its odds are 99:1 or 2:1, as any bookmaker will readily tell you.

If we denote as $\pi$ the probability of a certain event happening, the modeling assumption for logistic regression is that

$$Y = \log \frac{\pi}{1-\pi} \tag{5.19}$$

is a linear function of the predictors $\mathbf{X}$, or

$$\log \frac{\pi}{1-\pi} = \mathbf{X} \cdot \beta, \tag{5.20}$$

where once again we have added a constant feature equal to 1 (a bias) to our feature matrix **X**, so the intercept is absorbed in the matrix multiplication. The transformation expressed by Eq. 5.19 is called *logistic unit*, or *logit* in short form.

The inverse transformation connects Y, the linear function of the predictors X, to $\pi$, the probability that we would like to predict:

$$\pi = \frac{1}{e^{-\mathbf{X}\cdot\beta} + 1}. \tag{5.21}$$

The transformation above is an example of *sigmoid* function, which will return with a vengeance in Chapter 8, when we talk about building blocks of neural networks.

Keep in mind that even if logistic regression can, possibly, provide a suitable model for predicting odds-type outcomes, there is still a strong modeling choice at its core, which is the assumption of a linear relationship between the auxiliary variable $Y$ (the logit function of the probability) and the predictors $X$. If this assumption is not valid or justified, the whole model won't have predictive power, just as was the case with linear regression. In other words, not all odds-type outcomes can be successfully modeled using logistic regression.

### 5.7.2 Loss function for logistic regression

Just as for any regression model, we should choose a loss function before building a model. In principle, we can freely choose what we like, but if we were approaching the problem from a probabilistic inference perspective, then only one metric would be associated with the maximum likelihood estimator, so we will follow this path here as it allows us to introduce a popular class of loss functions.

If our features are represented by vectors of predictors $X_i$, and our target is the vector of observed classes $y_i$ (let's say $0/1$), then the likelihood of observing each data point is the product of the probability $\pi$ of observing the class $y_i$ times the probability $(1-\pi)$ of observing the class $(1-y_i)$, multiplied for all the data points, since we assume that the measurements are independent:

$$\mathcal{L} = \Pi_{i=1}^{n} \pi_i^{y_i} (1-\pi_i)^{1-y_i}. \tag{5.22}$$

The logarithmic likelihood becomes a sum:

$$\log \mathcal{L} = \sum_{i=1}^{n} y_i \log(\pi_i) + (1-y_i) \log(1-\pi_i). \tag{5.23}$$

Our loss function for logistic regression will simply be the negative of the expression above, and is called *cross-entropy* loss, or *logistic loss*, sometimes abbreviated as log-loss.

We can further manipulate this expression by collecting the $y_i$:

$$\log \mathcal{L} = \sum_{i=1}^{n} y_i \log(\frac{\pi_i}{1-\pi_i}) + \log(1-\pi_i). \tag{5.24}$$

The variable appearing in the first log is exactly our auxiliary function from Eq. 5.19, so we can write it as $\mathbf{X}_i \cdot \beta$; if we want to make explicit the dependence on the coefficients (or weights) $\beta$, we obtain

$$\log \mathcal{L}(\beta) = \sum_{i=1}^{n} y_i(\mathbf{X}_i \cdot \beta) - \log(1 + e^{(\mathbf{X}_i \cdot \beta)}). \tag{5.25}$$

Unlike the linear regression case, there is no closed-form solution for the weights $\beta$, but the minimum of the loss function (i.e., the maximum likelihood) can be found numerically by using the gradient descent techniques introduced earlier in this chapter.

### 5.7.3 Detecting phase transitions with logistic regression

The first example we consider is drawn from the helpful review of [Mehta et al., 2019], which contains several other examples of usage of GLMs in statistical physics and beyond.

In this example, we use logistic regression to classify the states of the two-dimensional Ising model [Ising, 1925] according to their phase of matter. In the model, a material is represented by a lattice of atoms; each of them can have positive or negative magnetic spin. The model connects the degree of alignment between spins with the magnetic behavior of the material. I loved this blog post[4] for an introduction to Markov Chain Monte Carlo methods to simulate states in the Ising model.

The Hamiltonian for the classical Ising model is given by

$$H = J \sum_{ij} S_i S_j, \tag{5.26}$$

where the $S_i$ are the spins of each particle on the lattice and can have values $+1$ or $-1$, and $J$ is an interaction term. The system undergoes a phase transition (from an ordered ferromagnet with all spins aligned to a disordered phase) around a critical temperature $T_c$. In the thermodynamic limit (i.e., for infinite particles, infinite volume, and fixed density), it is possible to calculate the critical temperature analytically: $T_c/J = 2/log(1+\sqrt{2}) \sim 2.26$. For any finite system size, this critical point is

4 https://tanyaschlusser.github.io/posts/mcmc-and-the-ising-model/

 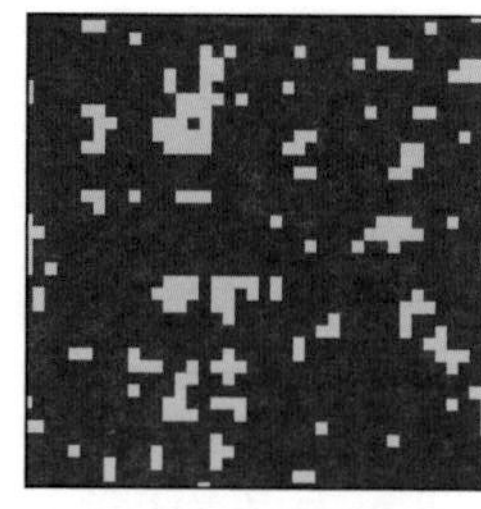 

**Figure 5.6: Examples of typical states of the two-dimensional Ising model for three different temperatures in the ordered phase (T/J = 0.75, left), the critical region (T/J = 2.25, middle) and the disordered phase (T/J = 4.0, right). Adapted from [Mehta et al., 2019].**

smeared out to a critical region around $T_c$. An interesting question to ask is whether one can train a statistical classifier to distinguish between the two phases of the Ising model. If successful, this strategy could be used to find the critical point in more complicated models where an exact analytical solution has so far remained elusive (e.g., [Morningstar and Melko, 2018]).

We use the simulated data from [Mehta et al., 2019], which describe the spin positions of a $40\times40$ lattice for different ratios of Ising interaction by temperature: $J/T \in (0.25, 4)$. The features of our logistic regression model are the values of the spins at every location, for a total of 1,600 features; the dependent variable is the class (in this example, 0 for disordered, 1 for ordered). The model derives the class by modeling the log-odds that a spin configuration belongs to the disordered class with a linear model.

Before starting, we have learned to ask: Should we expect this model to work well? After all, we know that just because logistic regression *can* provide a good model for odds-type problems, there is no guarantee that this will happen. By looking at a few examples of spin configurations, as shown in Figure 5.6, we can conclude that a good classifier will be sensitive to the degree of *agreement* between features, rather than their isolated numerical value. Furthermore, this classification problem has a "cyclic" behavior, so that configurations where every spin value is $+1$ are equivalent, class-wise, to configurations where every spin value is $-1$. Because linear models or generalized linear models, such as logistic regression, can't take into account interaction between features, it is hard to imagine that we will obtain very accurate results.

Loading a logistic regression model in `sklearn` is quite simple:

```
from sklearn.linear_model import LogisticRegression
model = LogisticRegression()
```

One thing worthy of notice is that Ridge regularization is automatically applied to the logistic regression model in `sklearn`, so that the standard loss function has the form

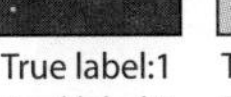

**Figure 5.7: Result of logistic regression classifier on eight test examples. The model fails to correctly classify examples 3 and 7, which are predicted to be in the ordered phase with high confidence, while to the human eye, they are easily recognizable as members of the disordered class.**

$$\log \mathcal{L} = C \sum_{i=1}^{n} y_i \log(\pi_i) + (1 - y_i)\log(1 - \pi_i) + ||\beta||_2^2 \tag{5.27}$$

(the actual expression given in the documentation is slightly different because of a difference in the reference value for the positive/negative class; here we assume it is $1/0$, while in `sklearn` they are automatically encoded as $1/-1$; see, e.g., this answer[5]).

The coefficient $C$ controls the amount of regularization; for large $C$ values, no regularization is applied, because the term that controls the norm of the weights becomes negligible. The default value is $C = 1$, but the best values should always be chosen through CV, especially if the data are not standardized beforehand, so there is no natural scale parameter to set the balance between the two terms.

Other parameters that can be tuned include the type of regularization (L1 or Lasso, L2 or Ridge, Elastic-Net) and the solver used to find the miminum of the loss function in the gradient descent procedure (liblinear, SGD, etc.).

In our case, as shown in the notebook "LogisticRegression.ipynb," the highest classification score is for $C = 1.0$ and above, but it remains quite low at around 68%, which means that 1 out of 3 examples are misclassified. The model also exhibits overfitting, with a significant difference between train and test scores.

Some examples of classification are shown for eight test configurations are shown in Figure 5.7. While the model is successful in classifying all the examples in the "very ordered" configuration, including some with mostly $-1$ spins (examples 2 and 6) and some with mostly $+1$ spins (examples 1 and 5), as well as two of the "disordered" examples (4 and 8), it fails to recognize examples 3 and 7 as belonging to the "disordered" class. In fact, it assigns them to the ordered class *with high probability* (95 and 62%, respectively). This example highlights the limitations of this model in recognizing classes.

In the exercise section, you are invited to consider possible improvements (by selecting a different algorithm and through feature engineering) to the logistic regression model.

5 https://stats.stackexchange.com/questions/468356/logistic-regression-loss-function-scikit-learn-vs-glmnet

## 5.8 POISSON REGRESSION

In this section, we consider a different type of GLM, the so-called *Poisson regression*. We mentioned already that the ingredients of a GLM are a random variable described by a function in the exponential family (Gaussian, gamma, Poisson, binomial,. . .), a link function, and linear predictors (the features, in ML-friendly language). The modeling assumption is that the transformed response variable, using the link function, is linear in the predictors. In the logistic regression example discussed in the previous section, the link function was the *logit*, and the response variable (the probability of belonging to one of two classes) was modeled by the binomial distribution. Here, we explore a different combination.

We consider a problem from [Vandenberg-Rodes et al., 2016]: We attempt to predict the number of hours that *nuisance flooding* will occur at the locations of 18 tide gauges along the coasts of the United States, as a function of mean sea level during the same year.

The interest of this problem for geophysicists is the ability to predict the impact that the observed and forecast rise in sea level due to global warming will have on the global economy through nuisance flooding (as opposed to destructive flooding, which is more damaging but also less frequent). Examples of negative socioeconomic impacts from nuisance flooding include damaging infrastructure (i.e., surface transportation and sewer systems) and posing public health risks.

The data that we use are from the National Oceanic and Atmospheric Association; see [Vandenberg-Rodes et al., 2016] and the associated repository[6] for details. Our only feature is the yearly mean sea level at the location of interest. The response variable (i.e., the nuisance flooding, NF) is defined as the cumulative number of hours that the water level exceeds a certain threshold, in each meteorological year. Our sample is composed of yearly historical measurements, which are typically available from the 1920s on for most of the tide gauges. The geographic distribution of the tide gauges is shown in Figure 5.8.

What framework shall we use to describe our response variable, NF? Because we are modeling a number of hours, which is a type of *count*, we can attempt to use the Poisson or negative binomial distributions, which describe nonnegative integer variables. For the sake of simplicity, we will assume here that the NF is well described by a Poisson distribution (but you can check the exercise section and the original paper for other options).

The link function in this model is the *logarithm* function; this makes sense, because the response is nonnegative, so our linear model is

$$\log \mu = \beta_0 + \beta_1 \cdot \mathrm{MSL}, \tag{5.28}$$

6 https://bitbucket.org/vandenbe/sea_level_final/src/master/

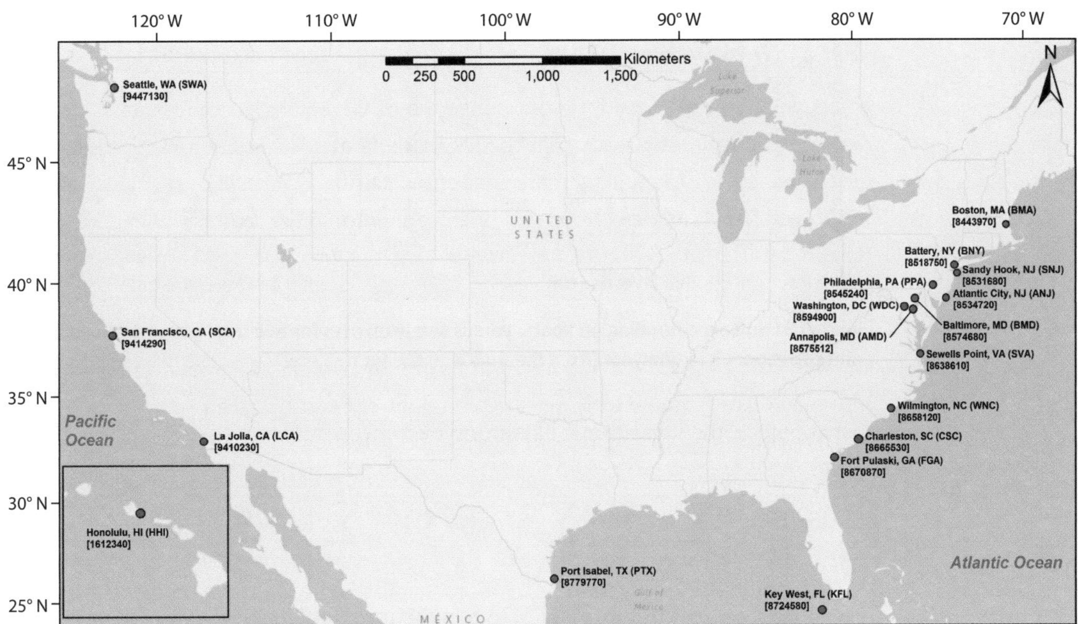


**Figure 5.8: Locations of the 18 tide gauges. From [Vandenberg-Rodes et al., 2016].**

where $\mu$ is the expected value of the response variable (NF), and MSL is the mean sea level. Note that this is a two-dimensional problem, so $\beta_0$ and $\beta_1$ are scalars.

In the GLM framework, the choice of exponential family also conditions the choice of the error model: In the Poisson distribution, the variance is equal to the mean, so this generates a naturally heteroskedastic model with larger dispersions associated with larger measured values. This assumption might not be perfect but is at least sensible in this case, as we can see by plotting the measured data for a couple of stations and looking at the dispersion of the NF variable across a range of values, as shown in Figure 5.9. Additionally, as we know, in the ML framework, we can afford to be less rigorous, because we won't claim to build a probabilistic inference approach, although we still follow the conventions of the statistical model when choosing our loss function.

Before deploying the algorithm, we have to select a loss function. For linear regression, we were able to write the MSE loss and derive the coefficients that minimize it directly, through the OLS approach. For the logistic regression, we started from the expression of the likelihood and derived the loss function as $\log \mathcal{L}$. Here, we follow a similar path. The likelihood of observing counts $y_i$ for our data, given a feature vector $\mathbf{X_i}$, is given by

$$\mathcal{L} = \Pi_{i=1}^{n} \frac{e^{-\mu(\mathbf{X_i})}[\mu(\mathbf{X_i})]^{y_i}}{y_i!}. \tag{5.29}$$

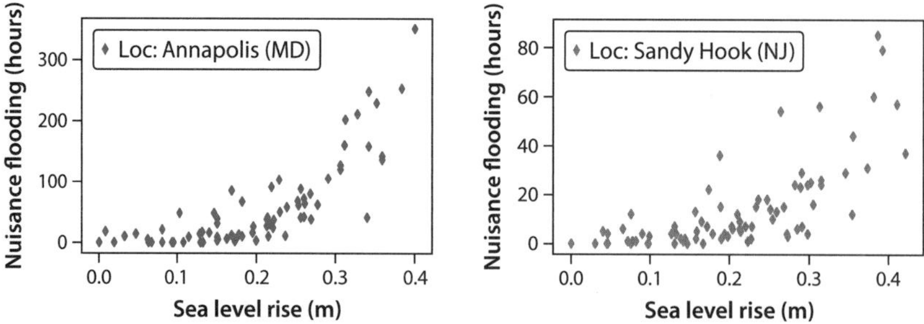


**Figure 5.9: Nuisance flooding, in hours, versus sea level rise, for two tidal gauge stations. Each point corresponds to one year (from the 1920s to today).**

Just as before, the logarithmic likelihood becomes a sum:

$$\log \mathcal{L} = -\sum_{i=1}^{n} e^{\mu(\mathbf{X_i})} + \sum_{i=1}^{n} y_i \log(\mu(\mathbf{X_i})) - \sum_{i=1}^{n} \log(y_i!). \tag{5.30}$$

Finally, we can use formula 5.28 to substitute $\log(\mu(\mathbf{X_i})) = \beta_0 + \beta_1 x_i$, where our only feature $x_i$ is the mean sea level (MSL):

$$\log \mathcal{L} = -\sum_{i=1}^{n} e^{(\beta_0 + \beta_1 x_i)} + \sum_{i=1}^{n} y_i(\beta_0 + \beta_1 x_i) - \sum_{i=1}^{n} \log(y_i!). \tag{5.31}$$

In GLMs, it is conventional to report not the log-likelihood, but a slightly different quantity called *deviance*; this is the case for both libraries we consider below. The deviance of a model with parameters $(\beta_0, \beta)$ is defined as $2\log \mathcal{L}(\text{saturated}) - \log \mathcal{L}(\beta_0, \beta)$, where the first term is the log-likelihood of the saturated model, which has one parameter per data point. For the Poisson regression, $\log \mathcal{L}(\text{saturated}) = 0$, so the deviance is just twice the log-likelihood.

We can now implement a Poisson regression model (without regularization) for a few tide gauge stations. In Python, this can be done using (at least) two different libraries: `statsmodels`, which is based on the R-language equivalent, and as of May 2020, `sklearn`. The latter has the familiar signature:

```
from sklearn.linear_model import PoissonRegression
model = PoissonRegression()
```

The lecture notebook "NuisanceFlooding.ipynb" shows both approaches, which lead to identical results.

Figure 5.10 shows the results of applying a linear regression model and a Poisson regression model to the data for the Annapolis tide gauge station, after eliminating data points for which measurements were not available. The linear regression model obviously fails to have any predictive power, while the Poisson model is at least sensible. Two possible ways to quantify such differences are to compare the

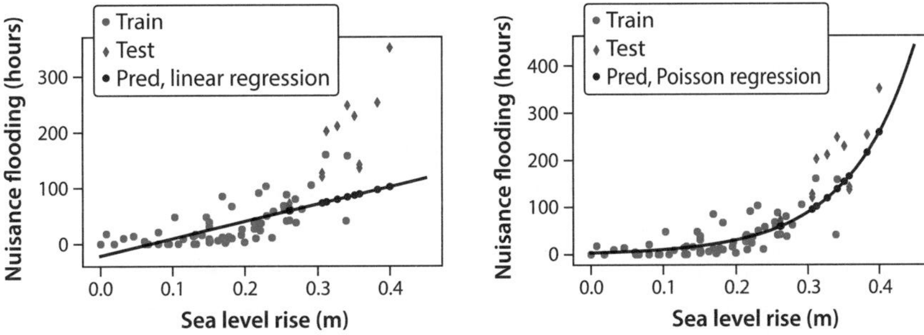


**Figure 5.10: Predicted (black) vs true (blue) nuisance flooding values for the linear regression model (left) and the Poisson regression model (right) for the Annapolis tidal gauge station.**

MAE (used in the paper [Vandenberg-Rodes et al., 2016]) and the $R^2$ score of the two models. For the linear regression, the MAE of predicted vs true test values is 99.0 hours, and the $R^2$ score is $-1.27$; a negative value indicates that the model performs worse than a model that predicts a constant value equal to the mean of the $y$ values in the test set. For the Poisson regression model, those values are 53.0 hours and 0.36, respectively, a significant improvement in both scores.

## 5.9 LESSONS LEARNED

In this chapter, we switched focus from classification to regression problems. Because we are still under the umbrella of supervised learning, a lot of what we learned about data exploration and preparation, diagnostic tools, and hyperparameter optimization remains relevant and valid. However, regression models required us to think about a whole new set of evaluation metrics, and we learned about a new class of algorithms, linear models:

- We introduced and discussed different metrics for regression, including the coefficient of determination $R^2$, the MSE, the mean absolute error, and the Huber loss.
- We investigated how linear models can be applied in an ML context and discussed the relation to the probabilistic inference approach. We also discussed the analytic solution of the linear model, known as the normal equation.
- We discussed a numerical solver for optimization problems, gradient descent, which is significantly faster and more flexible than the normal equation, as it can work with any loss function. We introduced different flavors of gradient descent such as SGD and mini-batch gradient descent, which are able to further improve time complexity.
- We learned about the role of regularization in improving the robustness of regression models and placed it in the (evolving) context of the bias-variance trade-off framework.

- We discussed two regularization strategies, Ridge and Lasso regularization, and highlighted the differences between them, in particular, the tendency of Lasso to induce sparse cofficients, which makes it suitable for dimensionality reduction/feature selection, as discussed in the following chapters.
- We considered a more general class of linear models, which expand our modeling capabilities while retaining the simple machinery associated with linear models, and discussed an example of logistic regression and one example of Poisson regression.

A final note on linear models is that, as was shown by GLMs, they can be used to model much more flexible relationships than their name suggests! For example, imagine using the `PolynomialFeatures` utility function shown in the lecture notebook "Regularization.ipynb" to generate a bunch of polynomial combinations of your original features and then applying a linear model in your new feature space. Is this still a linear model? Definitely! A linear model only requires that the target is linear in the predictors (a.k.a. the features), but it most certainly doesn't care about whether you are using some complicated function of your original inputs. Therefore, with a bit of preprocessing, you can use the desirable framework of linear models, which includes a (possibly regularized) analytic solution, flexible regularization options, and high interpretability, to model a wider range of input/output relationships.

## 5.10 REVIEW AND DISCUSSION QUESTIONS

Note: Questions and exercises marked by ** are more complex, open-ended, or time consuming. Those marked by * have more than one correct answer.

**Exercise 1.** What is the best definition of a linear regression model?

A) The model creates a linear relationship between input and output with fixed coefficients.
B) The model creates a linear relationship between input and output and finds the best coefficients (e.g., slope and intercept in two dimensions).
C) The model creates a linear relationship between the input variables and the output, but the output can only be 0 or 1.
D) The model can mimic any form of relationship between input and output.

**Exercise 2.** Which model, linear or logistic regression, would be best to fit the following problem: "Find the relationship between age and probability of owning a car"? Why? Can you discuss some limitations of your model?

What other variables (features) would be important to consider for the above problem?

**Exercise 3.** The coefficient of determination $R^2$ is used to. . .

A) Evaluate how good a regression model is.
B) Calculate the accuracy in a classification model.
C) Calculate the ratio between slope and intercept in a linear regression model.
D) Estimate on what fraction of the data the model fails.

**Exercise 4.** If your data are scattered all over a plane and you fit them with a straight horizontal line at the height of the mean *y* coordinate of the data, you expect your $R^2$ score to be. . .

A) Negative
B) Positive
C) Zero

**Exercise 5.** Why can't you use metrics such as accuracy or F1 score in a regression problem?

**Exercise 6.** * What can gradient descent be used for? Select all that apply.

A) To find the coefficient of determination
B) To solve iteratively for the coefficients of a linear regression problem
C) To change kernels in a Support Vector Machine
D) To minimize functions

**Exercise 7.** To solve a linear regression problem with MSE loss for a data set with $n$ examples and $m$ features, if $n$ is large and $m$ is small, which of these methods would you NOT use?

A) The normal equation
B) Batch gradient descent
C) Stochastic gradient descent
D) Mini-batch gradient descent

**Exercise 8.** To solve a linear regression problem with MSE loss for a data set with $n$ examples and $m$ features, if $n$ is small and $m$ is large, which of these methods would you NOT use?

A) The normal equation
B) Batch gradient descent
C) Stochastic gradient descent
D) Mini-batch gradient descent

**Exercise 9.** * Regularization can be employed in ML models as a way to. . . (select all that apply)

A) Reduce overfitting
B) Reduce underfitting
C) Select important features
D) Build additional features

## 5.11 PROGRAMMING EXERCISES

**Exercise 1.** ** Start from the lecture notebook about gradient descent ("LinearRegression_Gradient Descent.ipynb"), and note what happens for larger learning rates and smaller learning rates. Would an adaptive learning rate be a solution? Qualitatively, how would you choose it?

**Exercise 2.** In the same notebook as in the previous exercise, examine the gradients to discover why batch GD stops updating the slope pretty quickly. For loss functions that are not convex, would this be a concern in terms of getting stuck in local minima?

**Exercise 3.** ** Write your own CV routine for Ridge and/or Lasso regression, and compare it to LassoCV and RidgeCV in `sklearn`.

**Exercise 4.** ** Write a gradient descent algorithm for a loss function of your choice; compare with the results obtained using a grid-search approach.

**Exercise 5.** ** After reviewing the "LogisticRegression.ipynb" notebook, discuss the pros and cons (expected performance, timing, etc.) of the following algorithms: kNN, Decision Tree, and SVM, for the Ising model data set.

**Exercise 6.** ** Implement a method of your choice among those mentioned in the previous exercise and check your intuition.

**Exercise 7.** Reduce the dimensionality of the lattice from 40 to 20 by combining togetheradjacent cells and rounding the average spin (e.g., a cell with 4 pixels with spins 0,1,1,1 would have spin 1, a cell with 0,0,0,1 would have spin 0). Incidentally, this approach will be helpful for understanding pooling in neural networks. Use SVMs on the original and on this new data set, and find the performance of the optimal model.

**Exercise 8.** ** Can you engineer some features that might improve the performance of the logistic regressor in the above example?

**Exercise 9.** ** Using the data from the nuisance flooding example from Section 5.8,experiment with a different GLM model. 1. What distributions from the exponential families would be suitable for this problem? 2. Are you able to reproduce the paper's results using a negative binomial distribution? 3. Are you able to beat the performance of the Poisson regressor model quoted in the paper, using the MAE as your evaluation metric?

**Exercise 10.** ** Implement a logistic regression classification algorithm for the occurrence of solar flares, using the data from [Florios et al., 2018] and available here.[7] Can you match the performance (accuracy) reported in the paper?

7 https://data.mendeley.com/datasets/4f6z2gf5d6/1

CHAPTER SIX

# Ensemble Methods

*Where can you find indecisive hikers?*
*In a random forest.*

Our journey so far has taken us through several algorithms that can be used for classification and regression problems; we have learned to optimize each one of them and to compare different models.

In this chapter, we explore a different technique to improve the performance of single models, which consists of considering the aggregate response of several models instead. From a practical perspective, we could imagine training a bunch of classifiers or regressors and then using the most frequently predicted class or the average of the regression predictions for our final model.

Before proceeding, we should probably ask: Why would we want to do such a thing? The practical answer is that, in many cases, the combined model is more accurate than any of its constituents.

Let us consider the example of binary classification. Imagine that we have trained $N$ classifiers, each with an accuracy rate of $p$ (i.e., returning the correct outcome $p$% of the time). If the models are independent of one another (i.e., their outcome is uncorrelated), the probability that $k$ classifiers are correct is given by the binomial distribution:

$$P(k, N, p) = \binom{N}{k} p^k (1-p)^{N-k}. \tag{6.1}$$

If we simply assume that the classification of the ensemble model is given by a majority vote, the combined model will be correct if at least half of the individual classifiers are correct. More precisely, the probability of the combined model to be correct is

$$p(\text{comb model}) = \sum_{i=N/2}^{N} \binom{N}{i} p^i (1-p)^{N-i}, \tag{6.2}$$

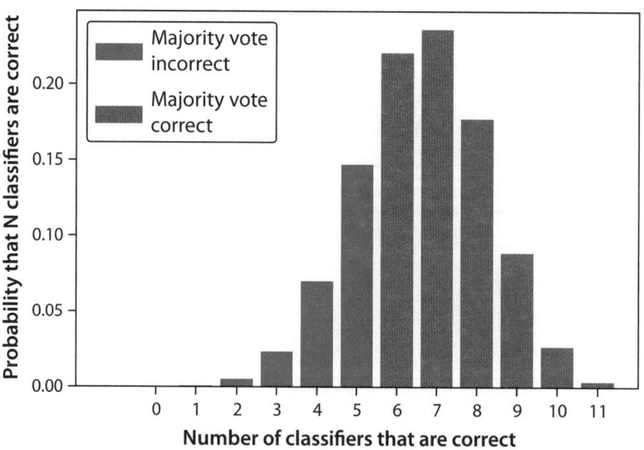


**Figure 6.1: The histogram shows the probability that N classifiers are correct out of an ensemble of 11, if the individual accuracy of each classifier is 60%. The majority-vote classifier will be correct if at least 6 models in the ensemble are correct; the probability of that is 75%, which is significantly higher than the single-classifier accuracy.**

which will be larger than the individual classifier's success rate, $p$, as long as $p > 0.5$ (i.e., each individual classifier is better than random).

For a concrete example, let's have $N = 11$ classifiers, so we don't have to break ties, each one with a rate of success of 60%. The probability that $1, 2, 3, \ldots, 11$ of them are correct is shown in Figure 6.1, and the ensemble model, which votes like the majority, is correct if at least 6 of them are correct. Summing up the probabilities that $6, 7, 8, \ldots, 11$ of them are correct, we obtain an accuracy for the combined estimator of 75%, which is quite significantly higher than the accuracy of each individual classifier.

[Dietterich, 2000], from which the above example is inspired, goes further in listing three practical reasons that an ensemble model can perform better than a single model.

The first reason is *statistical*: Because of the finite size of the training set, it is possible to find several I/O relationships that perform equally well on the training data; taking an average of them tends to shift us toward the "true" relationship. The second reason is *computational*. Many learning algorithms work by performing some form of local search for the minimum of the loss function, for example, by gradient descent, and they may get stuck in local optima. An ensemble constructed by running the local search from many different starting points may provide a better approximation to the true unknown I/O relationship. Finally, the third reason is *representational*: in many cases, the true I/O relationship cannot be represented by any of the available hypotheses. However, ensembling (forming weighted sums of hypotheses) expands the space of functions that can be represented, allowing us to explore more flexible relationships.

## 6.1 BIAS-VARIANCE DECOMPOSITION FOR ENSEMBLES

Much of the theory behind ensemble methods is based on the idea of bias-variance decomposition, which was introduced in Chapter 5. The idea is that for the mean squared error (MSE) loss, it is possible to decompose the generalization error (i.e., the error on previously unseen data) of model $\phi$ at feature value $\mathbf{x}$ into three pieces:

$$E[\mathrm{MSE}(\phi(\mathbf{x}))] = \mathrm{noise}(\mathbf{x}) + \mathrm{bias}^2(\mathbf{x}) + \mathrm{var}(\mathbf{x}). \tag{6.3}$$

As a reminder, the first term is an irreducible noise term that is independent of the learning algorithm; in statistical learning, this is the error of the Bayes model, the best-performing possible model given a certain loss function. The second term, the bias, is the average deviation of the prediction from the true answer; and the third term, the variance, measures the variability of the predictions at coordinate $\mathbf{x}$ over the models learned from all possible learning sets [Louppe, 2014]. Note that the validity of the bias-variance *decomposition* shown above is independent of the controversial bias-variance *trade-off* also introduced in the previous chapter, which states that the bias will decrease and the variance will increase as a function of increasing model complexity.

The actual balance of the decomposition is different for each algorithm, but we will see in the following sections that the process of ensembling can be used to reduce either variance or bias.

In many modern ensemble methods, the base estimator is a decision tree; this is convenient, because decision trees are quite flexible, and they can be used in a parallel fashion using fully developed trees or in a sequential fashion by means of one- or two-node trees, often called *stumps*. However, ensemble methods can be realized using different base learners. We will analyze some popular choices in the following sections, after describing the data set and problem to be used.

## 6.2 A THREE-DIMENSIONAL MAP OF THE UNIVERSE

The problem considered in this chapter is an important one in astronomy: the determination of the so-called photometric redshifts for galaxies.

When we look at the night sky, either with the naked eye or with the help of a telescope, objects appear to be on a two-dimensional sphere around us, what the ancient Greeks used to call *the celestial sphere*. However, we now know that the various astronomical objects that we see, such as stars and galaxies, are located at different distances, which might be as large as several billion light years, from us. Measuring these distances is very important, for many reasons; for example, it allows us to create a three-dimensional map of the Universe, and it allows us to turn measurable apparent brightnesses (which scale with inverse-square distance) into intrinsic luminosities, thus obtaining information about the true energy emission of astrophysical sources. Furthermore, because light travels at a finite and constant speed, a distance can be interpreted as a lookback time; we can therefore reconstruct the evolutionary history of the Universe by observing sources in "time slices."

But how can we measure the distances of objects such as galaxies, given that direct measurements are precluded by the enormous distances? The key element for doing this is the law of expansion of the Universe, colloquially known as the *Hubble Law*. It states that *the farther away a galaxy is, the faster it is moving away from us*. The spectra of all galaxies that are sufficiently far away to not get caught in local gravitational fields (i.e., that are in the *Hubble flow*) are observed at lower frequencies compared to similar ones in the local Universe: This is essentially a Doppler shift, observed on light instead of sound, of cosmological origin. Because in the visible part of the electromagnetic spectrum, red represents the lower-frequency end, this phenomenon is called *redshift*. All spectra of distant galaxies are redshifted compared to the original light emission, and the redshift $z$ can be expressed directly as a stretch in the observed wavelength with respect to the emitted one as

$$1+z=\frac{\lambda_{obs}}{\lambda_{\rm em}}. \tag{6.4}$$

The difference between $\lambda_{obs}$ and $\lambda_{\rm em}$ is caused by the expansion of the Universe and can be interpreted in terms of a scale factor $a$, which describes the linear stretching of spacetime and is typically normalized to be equal to 1 at the present time:

$$1+z=\frac{a_{\rm obs}}{a_{\rm em}}=\frac{1}{a_{\rm em}}. \tag{6.5}$$

Just as for the Doppler effect, redshift is related to the receding velocity of galaxies, not strictly to their distance; but the Hubble law tells us that these two quantities are proportional to each other. We can actually be more precise and use our current model of the Universe, known as ΛCDM model (with a cosmological constant and cold dark matter), to write a more general version of the Hubble law, the law of distance vs redshift. There are several definitions of distances in a Friedmann-Robertson-Walker universe, but picking the comoving distance as a reference, and assuming that we live in a flat universe, we obtain

$$d_c(z)=\frac{c}{H_0}\int_0^z \frac{dz'}{\sqrt{\Omega_m(1+z')^3+\Omega_r(1+z')^4+\Omega_\Lambda}}, \tag{6.6}$$

where $c$ is the speed of light, $H_0$ is the current slope of the speed/distance relationship, and $\Omega_m$, $\Omega_r$, $\Omega_\Lambda$ correspond to the fraction of matter/energy budget, relative to the total, in the form of matter, radiation, and cosmological constant.

Assuming that those fractions are known (which is, to be fair, only approximately true!), *we can calculate distance (or lookback time) if we know the redshift.* This is awesome, because redshift is relatively easy to measure from spectra of galaxies, as we will see in the next section.

### 6.2.1 Photometric vs spectroscopic redshifts

The spectrum of a galaxy is a chart of its light emission as a function of wavelength. It comes from the aggregate light of billions of stars, and it's affected by many complicated phenomena that occur during galaxy formation and evolution. The chemical composition of stars and its evolution, the process of stellar assembly, the presence and properties of cosmic dust, and interactions (e.g., mergers, stellar winds, gas inflows and outflows, just to name a few) all contribute to shape the observed spectrum. So how can we say that measuring redshift is relatively easy?

The key factor is that despite the great complication in the overall shape and details, there are at least some features of the spectrum that originate in simple physics, for example, the emission or absorption of photons in atoms like hydrogen or oxygen at certain wavelengths. Compared to the galaxy's average light emission, the fact that these photons are more or less likely to exist manifests itself in the form of "spikes" or "dips" in the spectrum, generally called *lines*. If the lines can be correctly identified as belonging to specific atoms, because their emission wavelength is exactly known from laboratory experiments, the redshift can be easily calculated through Eq. 6.5. Other recognizable features include "breaks" (sudden increases or drops) that are due to properties of stellar atmospheres. This method is known as *spectroscopic redshift*, and is based on the assumption that a *spectrum* (i.e., a detailed chart with enough resolution to measure these features with confidence) is available. An example spectrum, where emission and absorption features have been identified and marked, is shown in Figure 6.2.

Because knowing the distance to galaxies is so crucial for accurately mapping the three-dimensional Universe, astronomers would like to estimate redshifts also for lower-quality data, where direct detection of single sharp features is not possible. This is often the case for *photometric* (imaging) data, which are similar to photographs: In this case, the chart of luminosity vs wavelength is much more coarse, and we only have information about the *average* brightness of a galaxy over a range of wavelengths, known as a *band*. In typical broadband photometry, the width of bands can be on the order of a thousand Å, and sharp features are completely "washed away" by the averaging process. While spectroscopy is the gold standard for measuring redshifts, obtaining photometric data is much less time consuming, because photons are grouped in much wider bins on their arrival at the telescope, and therefore it takes a much shorter exposure time to achieve a certain signal to noise ratio, or SNR. Therefore, we can avail ourselves of photometric data for much larger samples of galaxies in the Universe, and upcoming surveys are slated to image a significant

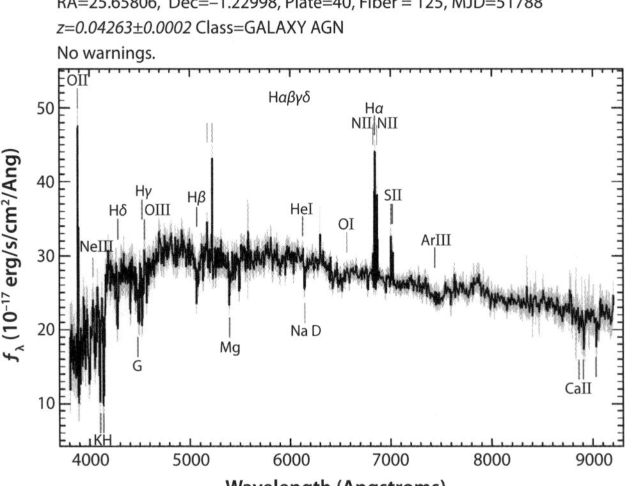


**Figure 6.2: Sharp features such as emission and absorption lines, if they can be correctly identified, help derive the spectroscopic redshift of galaxies. Spectrum from the Sloan Digital Sky Survey.**[1]

fraction (>10%) of all the galaxies in the Universe. Being able to derive reliable redshifts for these galaxies is crucial for astronomy.

The techniques that are used to estimate the so-called *photometric redshifts* are of two types (see [Salvato et al., 2019] for a recent review). The first is template based: We use linear superpositions of template shapes of known redshifts. The second, which is more relevant here, is through supervised learning and consists of using galaxies for which spectroscopic redshifts are known as a learning set.

## 6.2.2 Data

Our example is inspired by the work of [Zhou et al., 2019], who also made the data available here.[2] The goal of this exercise is to build a redshift forecasting algorithm for data resembling the wavelength coverage and depth of the upcoming Vera Rubin Observatory (formerly known as Large Synoptic Survey Telescope) [Ivezić et al., 2019], which is expected to provide photometry in six bands, ranging from near-ultraviolet to near-infrared (u, g, r, i, z, and y), of approximately 20 billion galaxies, spanning a considerable fraction of the Universe's volume.

Spectroscopic data used for training are from the DEEP2 [Newman et al., 2013], DEEP3 [Cooper et al., 2011], and 3D-HST [Brammer et al., 2012] surveys. We will attempt to solve this problem using different ensemble algorithms, as described in the next sections.

1 https://www.sdss.org/
2 http://d-scholarship.pitt.edu/36064

## 6.3 BAGGING METHODS

Bagging methods, which owe their name to Bootstrap AGGregating, are a popular set of models that rely on a two-step process: First, a re-sampling (bootstrapping) of the learning data generates some *randomized* learning sets, and second, an average (aggregating) of the responses of the individual models provides the ensemble prediction.

In such models, the assumption is that the individual learner has lower bias (is quite accurate on the training set) and higher variance than the optimal model (i.e., the model with the lowest test error). The ensembling procedure attempts to identify a different bias/variance trade-off point by obtaining a reduction in variance *larger than the increase in bias due to the randomization*, so that the total test error is lower.

Perhaps the most famous bagging algorithm is *Random Forests*, which consists of a randomized ensemble of decision trees. We focus on this algorithm and its close relative, the Extremely Randomized Trees, in this section.

### 6.3.1 Decision trees recap

We can start by recalling some details from the decision trees algorithm introduced in Chapter 3. In that example, we were using the algorithm for classification purposes, but the behavior is quite similar for regression. The main idea is that at each step, we consider all possible splits on all possible features and choose the one that leads to the highest reduction in impurity of the resulting branches. For classification, one common metric used to assess the quality of the splits is the Gini impurity, while in regression problems, it is common to use the mean squared error or mean absolute error. The process continues until some convergence criterion is reached. In classification problems, this will happen when all leaves are pure (every single object in the training set has been classified correctly). In regression problems such as the one considered here, we can never reach an "exact" agreement between predicted and true values, and therefore the maximally deep tree will have one object in each of its leaves, so that no further splits are possible. As we learned, maximally developed trees have a strong tendency to overfit, so it is common to limit the depth of trees by tuning some of the parameters described in the following section.

In the beautiful language of pseudocode, decision trees can be summarized as follows (inspired by [Louppe, 2014]):

```
function DecisionTreeRegressor(L)
if the stopping criterion is met for L then
     predict ŷ = constant
else
     Find the split on L that maximizes decrease of impurity:
     s* = arg max Δi (s)
```

```
        Partition L into L_L and L_R according to s*
        add new node = DecisionTreeRegressor(L_L)
        add new node = DecisionTreeRegressor(L_R)
end if
```

#### *6.3.1.1 Hyperparameters of decision trees*

Tunable hyperparameters of decision trees are mostly related to balancing the higher degree of purity afforded by additional splits with the higher risk of overfitting incurred by more developed trees. The procedure of reducing the complexity of a fully developed tree is referred to as *pruning*. There are a few hyperparameters that act in a somewhat similar manner:

- The minimum value of impurity decrease required to consider splitting a node (`min_impurity_decrease` in `sklearn` parlance), which can be set to a value larger than zero to decide whether a split is worth the additional complexity;
- The minimum number of samples required to consider splitting a node (`min_samples_split`); the absolute minimum, of course, is 2, but one can require a higher number to avoid the presence of very small leaves;
- The minimum number of samples required in a leaf node (`min_samples_leaf`); the absolute minimum, of course, is 1; this parameter is very similar to the one above; and
- The maximum depth (`max_depth`), which controls the overall number of splits allowed in each parallel branch of the tree. It acts similarly to the parameters above, but if the data set has nonuniform density, it allows for different-sized terminal leaves.

Typically, because of the overlap in meaning and effect of these parameters, we would only choose one or two to perform the optimization/tuning step.

### 6.3.2 Random Forests

As mentioned earlier, Random Forests are collections of randomized decision trees. In Random Forests, the randomization process occurs in two ways:

1. The original learning set is replicated several times using *bootstrap sampling with replacement*. If the size of the learning set is *N*, *N* elements are randomly picked from the learning set, one at a time (in other words, an element that has already been picked can be picked again and again). This creates *M* learning sets, on which *M* decision trees will be built.
2. When creating the *M* decision trees, the features that can participate in the selection of the optimal split are a random subset of the total features.

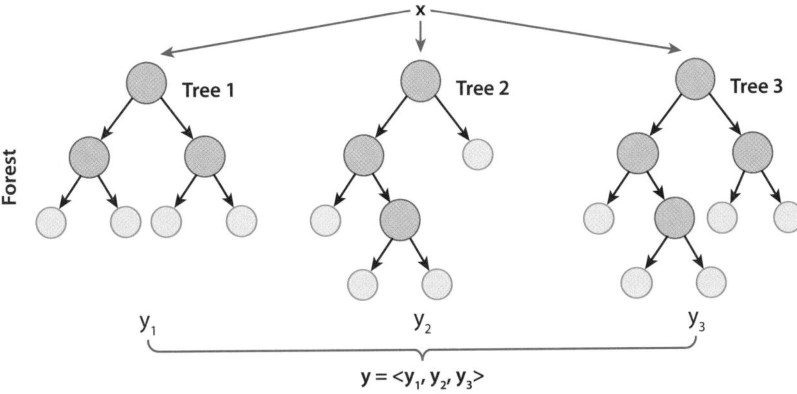


**Figure 6.3: The predictions of $M=3$ independent decision trees is averaged in a Random Forest. The trees differ from one another because they are built on different bootstrap iterations of the original learning set, and because a random selection of features participates in the selection of optimal splits at each step.**

The $M$ decision trees are built independently from one another, and each of them provides a prediction, $\hat{y}$, for each element in the learning set. The final prediction of the Random Forest is simply the mean of the predictions from the $M$ trees, as illustrated, for $M=3$, in Figure 6.3.

We have seen qualitatively why such collections can be expected to shift the bias/variance error decomposition, allowing for the possibility of lower generalization error, but here we can make a more specific argument.

Let us consider the decomposition of MSE loss in noise, bias, and variance, Eq. 6.3, and ask how it would change for an ensemble of randomized estimators (in this case, decision trees). We will make some considerations, and I refer you to [Louppe, 2014] and references therein for a more rigorous derivation of the second and third results.

- The noise term is independent of the algorithm used, and it will stay the same.
- It can be shown that the bias of an ensemble of randomized models is the same as the bias of any of the randomized models.
- The variance of an ensemble of $M$ randomized models can be decomposed as

$$\mathrm{var}(\mathbf{x}) = \rho(\mathbf{x})\,\sigma^2_{\mathcal{L}} + \frac{1-\rho(\mathbf{x})}{M}\,\sigma^2_{\mathcal{L}}, \tag{6.7}$$

  where $\rho$ is the correlation coefficient of the outcomes of two randomized versions of the same tree, which can be shown to be $\geq 0$; and $\sigma_{\mathcal{L}}$ is the typical variance of a single estimator on learning set $\mathcal{L}$.

The coefficient $\rho$ measures the strength of the randomization. If $\rho \to 1$, the variance of the ensemble is the same as that of a single tree. For higher degrees of randomization, however, $\rho < 1$, and it is possible to lower the variance, compared to the single tree, by using a large number of trees $M$. Of course, a higher randomization implies a higher bias in the single randomized tree, so we can't predict whether the total generalization error will decrease. However, *in situations where the generalization error is dominated by variance, using Random Forests of trees or other bagging algorithms is generally a promising tool to improve results.*

#### 6.3.2.1 Hyperparameters of Random Forests

The tunable hyperparameters of Random Forests include all those of the individual estimators, as well as a few related to the ensemble:

- The number of decision trees $M$ that compose the forest (`n_estimators` in `sklearn`). Note that adding more trees is always advantageous, because it decreases the second term in Eq. 6.7, but of course it increases training time. Typically, there is a plateau value for $M$, over which adding more trees does not result in additional benefits.
- The size of the random subset of features (`max_features`) used at each step; the smaller the subset, the higher the randomization.

### 6.3.3 Extra Random Trees

Extra Random Trees (ERTs) are very similar to Random Forests; the base estimators are decision trees, and the hyperparameters are very similar. The main difference between the two is the randomization process. In the `sklearn` implementation, the default choice is to have the following two randomization steps:

1. When creating the $M$ decision trees, the features that can participate in the selection of the optimal split are a random subset of the total features.
2. When creating the $M$ decision trees, random splits on each feature, as opposed to the best possible splits, are selected.

The first strategy is shared with Random Forests, but there is no bootstrap of the original learning set. Overall, the ERT algorithm tends to shift the bias/variance balance toward the higher bias/lower variance direction, compared to Random Forests, as a consequence of the enhanced randomization ensured by the second strategy.

## 6.4 BAGGING ALGORITHMS FOR PHOTOMETRIC REDSHIFTS

This section aims to essentially reproduce Fig. 6 of [Zhou et al., 2019]. You can follow along in the notebook "Photoz_RandomForests.ipynb." As usual, we make use of the `sklearn` API, and we can simply import the Random Forests and ERT algorithms from the "ensemble" package:

```
from sklearn.ensemble import RandomForestRegressor, ExtraTreesRegressor
```

It is worth noting that even though we are solving a regression problem, the corresponding functions and utilities for classification are also available.

We begin by selecting the RF algorithm with default parameters:

```
model = RandomForestRegressor()
```

As mentioned in Section 6.2.2, the features we will work with are the observed brightnesses of the galaxies, expressed in magnitudes, in six photometric bands, ranging from near-ultraviolet to near-infrared (u, g, r, i, z, and y). These are already present in the catalog (we use the "aper_cor" columns, as suggested in the discussion section of [Zhou et al., 2019]).

Our target property is the redshift of each galaxy, found in the "zhelio" column. The "true" values here correspond to the spectroscopically measured redshifts, although of course they are somewhat imperfect as well.

After loading the data, we can use our familiar cross validation (CV) technique to generate predictions and evaluate the results. This procedure generates an average $R^2$ score on the training set of 0.88, and average $R^2$ score on the test set of 0.34. We have discussed the lack of interpretability of the $R^2$ score in Chapter 5, but by focusing on the gap as opposed to the absolute result, we can easily see that we have a high variance problem: just what Random Forests were supposed to help us with!

To gain further insights, we can use the `cross_val_predict` function to generate predicted redshift values for a particular set of CV splits. Doing so produces the plot in the left panel of Figure 6.4. Unfortunately, the picture doesn't look at all like its counterpart in the [Zhou et al., 2019] paper! Other metrics that we can use for a fair comparison of results are the normalized median absolute deviation (NMAD) of the residuals vector and the fraction of outliers (OLF), defined as those objects for which the relative difference between true and predicted values is less than 15%:

$$\sigma_{\text{NMAD}} = 1.48 \times \text{median}\left(\frac{|\Delta z|}{1+z_{\text{true}}}\right), \tag{6.8}$$

$$\text{OLF} = \text{N}\left(\frac{|\Delta z|}{1+z_{\text{true}}} > 0.15 = T\right)/\text{N}. \tag{6.9}$$

Note that in both cases, the "relative" difference between predicted and true values is normalized by a factor $(1 + z_{\text{true}})$, instead of the usual $z_{\text{true}}$; this is to avoid weighing excessively the differences in the vicinity of $z_{\text{true}} = 0$. The 1.48 factor in the first equation is customary and is used to rescale the MAD to coincide with the standard

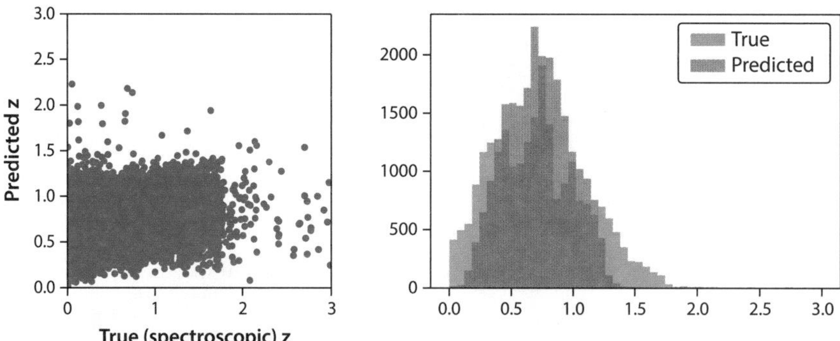


**Figure 6.4: Scatter plot (left) and distribution (right) of the true (spectroscopic) and predicted redshift values for the default setup described in Section 6.4.**

deviation for normal distributions. In our case, the normalized MAD is 0.057, and the OLF is a worrisome $\sim$ 25%: a far cry from the values of 1.8–3.8% and 4.5–6% reported in Figs. 5 and 6 of [Zhou et al., 2019].

One last useful bit of diagnostics/visualization that I like to explore is a plot of the *distributions* of true and predicted values, which are shown (for this particular CV splits realization) in the right panel of Figure 6.4. While such a graph does not provide the same object-by-object comparison of true and predicted values that the scatter plot shows, it is often useful for understanding how the algorithm is behaving *in general*. Here, note that the distribution of predicted values is "squeezed" with respect to the true one: Many galaxies are assigned redshifts closer to the mean of the distribution, and there are basically no predicted values at the edges of the distribution. We are encountering again one of the limitations of tree-based techniques mentioned when we first introduced decision trees in Chapter 3, the fact that the predicted distribution is narrower than the true one. Overall, it appears clear that we need to use some of the diagnostic tools and optimization techniques discussed in previous chapters.

### 6.4.1 Hyperparameter optimization

Our first attempt to improve performance relies on optimizing the parameters of the Random Forest. Because of the high variance issue, perhaps "pruning" the trees by limiting the total depth of each tree or ensuring a minimum size for each terminal leaf may be helpful. A similar result could be obtained by reducing the number of features that are randomly chosen to participate in splits. Additionally, we have learned that adding more base estimators (trees) might also help reduce the total variance, while leaving the bias unchanged. Therefore, one reasonable attempt at picking a parameter space could look like this:

- Vary the maximum depth of each tree (chosen values: 3, 6, and None);

- Vary the maximum number of features that participate in each tree's splits (chosen values: 2, 4, and None);
- Vary the minimum number of objects that are required in each terminal leaf (chosen values: 1, 5, and 10); and
- Vary the number of trees that participate in the forest (chosen values: 50, 100, and 200).

There is no hard-and-fast rule for choosing the range or values of parameters, because sensible numbers depend on the complexity of the data set and on the available computational resources. However, a good rule of thumb is to create an initial grid to get a sense of how the performance depends on each parameter, and to refine or expand the grid for parameters where a high gradient is observed. It is also helpful to include the benchmark model (the model with default parameters) in the grid: here, it is represented by the combination {`max_depth = None, max_features = None, min_split_leaf = 1, and n_estimators = 50`}.

An important "trick" to make optimization faster is to reduce the size of the data set before running the optimization. We know from learning curves that the behavior of both training and testing performance may vary significantly with data set size, so the "sweet spot" we are looking for is a size large enough that the train/test scores have plateaued, but significantly smaller than the original. In our case, I chose $N = 5{,}000$ and verified that the scores don't change significantly for the reference model, compared to using the whole data set.

The result of the cross-validated optimization for the top 8 models is shown in Figure 6.5. The parameter optimization does reduce the degree of overfitting (measured by the gap between train and test scores), but it only achieves a small improvement in the test scores. Furthermore, the test scores vary only modestly as a function of any of the considered parameters, suggesting that further parameter optimization is unlikely to have a significant effect. It looks like the secret of matching the paper's published performance lies elsewhere.

### 6.4.2 The importance of cleaning data

The previous sections are a good example of how important it is to understand the nooks and crannies of data selection and imputing, even when interpreting published results. Our approach seemed to be quite straightforward, but we failed to reproduce the results of the paper, because we didn't consider the selection criteria in detail.

In particular, there are a few quality flags and other data cuts that were used to select a *reliable* training set, and they were described in various sections of the reference paper:

- The redshift quality flag is $\geq 3$, meaning that the spectroscopic redshift measurement is highly reliable (and hence suitable to be used as "ground truth").

| Params | mean_test_score | std_test_score | mean_train_score |
|---|---|---|---|
| {'max_depth': None, 'max_features': 2, 'min_samples_leaf': 10, 'n_estimators': 200} | 0.362433 | 0.039260 | 0.521977 |
| {'max_depth': None, 'max_features': 2, 'min_samples_leaf': 5, 'n_estimators': 200} | 0.362081 | 0.040010 | 0.622369 |
| {'max_depth': None, 'max_features': 2, 'min_samples_leaf': 10, 'n_estimators': 50} | 0.361750 | 0.036359 | 0.520487 |
| {'max_depth': None, 'max_features': 2, 'min_samples_leaf': 5, 'n_estimators': 100} | 0.360990 | 0.039398 | 0.621199 |
| {'max_depth': None, 'max_features': 2, 'min_samples_leaf': 10, 'n_estimators': 100} | 0.360801 | 0.041837 | 0.521966 |
| {'max_depth': None, 'max_features': 4, 'min_samples_leaf': 10, 'n_estimators': 200} | 0.359517 | 0.040832 | 0.545785 |
| {'max_depth': None, 'max_features': 4, 'min_samples_leaf': 10, 'n_estimators': 100} | 0.358713 | 0.041134 | 0.546087 |
| {'max_depth': None, 'max_features': 4, 'min_samples_leaf': 5, 'n_estimators': 100} | 0.357760 | 0.040484 | 0.649448 |

**Figure 6.5: The results of cross-validated parameter optimization, shown for the top 8 models, for the setup discussed in Section 6.4.1. Despite a reduction in the degree of overfitting compared to the benchmark model with default parameters, the test scores remain unsatisfactory, and they don't exhibit a strong dependence on any of the parameters we explored.**

- Valid photometric measurements in all six bands are required; lack of observations or nondetections are characterized by the '99' or '−99' tag, while valid numbers are in the 20–25 range.
- For Fig. 6 of [Zhou et al., 2019], deep photometry from the Canada-France-Hawaii Telescope Legacy Survey (CFHTLS) data is required.

Performing these data cuts is quite simple using data frame slicing (also shown in the lecture notebook "Photoz_RandomForests.ipynb") and running the benchmark Random Forests algorithm with default parameters (and fixing the random seeds for reproducibility) reveals hugely improved average train scores with $R^2 = 0.96 \pm 0.01$, and test scores with $R^2 = 0.75 \pm 0.1$. Despite the lack of "absolute" interpretation of the $R^2$ score, it is still a useful tool to assess the relative improvement.

Because the scores still show high variance, we can run the same parameter optimization through a cross-validated grid search as before. In this case, the top 10 models have very similar scores to the reference model, indicating that the high variance issue is unlikely to be improved significantly by further tuning of the parameters. To compare our results with those of the paper more directly, we can calculate again the fraction of outliers and the NMAD, which come out to be 5.7% ($\pm$ 0.1) and 0.037 ($\pm$ 0.0004), respectively; the uncertainty is estimated by varying the random seeds.

These numbers should be directly comparable to those of Fig. 6 of the paper, because we have used the same selection criteria. While the OLF matches them quite closely, the NMAD is still about 50% larger than the published results. This discrepancy might appear puzzling, but a closer look reveals that there is one important residual difference between our methods and those applied in the paper. While the observational data are exactly the same, the authors engineered additional features by building combinations of magnitudes called *colors*. Colors measure the relative ratios of brightness in different bands and therefore are more sensitive to the shape of each galaxy's spectrum. They are built simply by taking the difference in magnitude in adjacent bands, (e.g., u-g, g-r, and so on); since the magnitude is a logarithmic measure of brightness, colors measure ratios of brightness. The paper uses five colors and the $i$ magnitude (which is helpful to retain, because it's a tracer for brightness), as features. The implementation of the Random Forests algorithm using colors is left as an exercise (but is included in the lecture notebook "Photoz_RandomForests.ipynb").

## 6.5 BOOSTING METHODS

Bagging methods like Random Forests or Extremely Random Trees work by building full trees in parallel and then averaging the predictions.

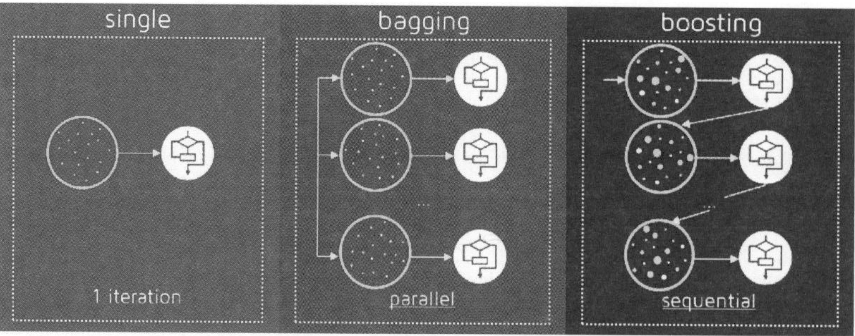


**Figure 6.6: A schematic illustration of ensembling by bagging (middle panel) and boosting (right panel) algorithms. Bagging methods consist of a population of fully independent estimators; boosting methods aim to create weak estimators in a sequential fashion. Figure reproduced with permission from Quantdare.com.**[3]

In contrast, boosting methods (e.g., [Kearns and Valiant, 1994; Schapire, 1990; Freund and Schapire, 1997; Friedman, 2001]) work *sequentially*: a weak, simple learner (in the case of tree-based methods, this is sometimes called a stump) is created and used to make predictions, then the next step focuses on getting the problematic examples right (i.e., "boosting" the success rate). This might happen by assigning higher weights to examples where the model is failing (as in the *AdaBoost* algorithm), or through updating the optimization function at every step (as in *gradient boosting machines*); we will consider these two popular classes of models in detail in the following sections. This procedure is applied iteratively until the performance is satisfactory. Unlike bagging methods, in which the elements of the ensemble are fully independent of one another, in boosting, the model is obtained by a series of progressive iterations. At each step, the performance obtained in the previous stage is used to determine the new solver, and the final model is a weighted additive combination of all the solvers. I liked this reference[4] for an informal description of boosting methods; a schematic summary of the differences between bagging and boosting algorithms is shown in Figure 6.6.

Just like the bagging ensemble algorithms presented above, we can use boosting methods both for classification and regression purposes. We begin by considering a toy example to illustrate how the decision-making process changes in the two methods and then show the boosting methods in action for determining photometric redshifts.

## 6.5.1 AdaBoost

The Adaptive Boosting (AdaBoost) method was first introduced for classification models by [Freund and Schapire, 1997]. In pseudocode, we can look at its main steps:

---

3 https://quantdare.com/what-is-the-difference-between-bagging-and-boosting/
4 https://explained.ai/gradient-boosting/index.html

1. Initialize weights $w_i$ uniformly (weights are $\propto$ to the probability $p_i$ of each sample to be part of the training set at any given stage).
2. Build, using sampling with replacement, a training sample of size N, using the weights above.
3. Build a regression model, i.e., a predictive hypothesis $h_t : x \rightarrow y$.
4. Calculate the average loss on training set; any loss function $L$ is valid, but it needs to be normalized so that $L \in [0,1]$.
5. Form the quantity $\beta = \frac{1-\bar{L}}{\bar{L}}$, which measures the degree of confidence in the predictor (low losses correspond to low $\beta$, and imply high confidence).
6. Update the weights: $w_i \rightarrow w_i \beta^{(1-L_i)}$. Examples with small losses (near-correct predictions) will incur a stronger weight reduction, and they are less likely to be picked as a member of the training set in the following iteration.
7. Break if $\bar{L} > 0.5$ (this corresponds to ensuring that $\beta < 1$); otherwise, repeat from 2. Note that because the system is given an increasingly difficult set of training examples, the average loss will increase!
8. Combine all the models using a weighted average or median of the models built at each iteration. The formulation by [Drucker, 1997], which is also used by sklearn, uses a weighted median with weights proportional to $log(1/\beta_t)$, where $\beta_t$ is the confidence measure at iteration $t$.

It is helpful to visualize the process of boosting and the role of the base estimators with a simple toy example with only one feature. We aim to model the shape of the function indicated by the training points in Figure 6.7, and show an example that uses a decision tree with a maximum depth of 3 and a decision tree with a maximum depth of 6. For both cases, we show how the model progresses with one, two, and three base estimators.

In the first case, shown in the left panel, it is easier to follow the behavior of the model. The decision pattern of the first estimator can be visualized using the same functions used in Chapter 2 when we first discussed decision trees; of particular importance is the first split, which happens at $x \sim 2.7$. The first split is chosen to minimize the combined MSE on the two sides of the splits; it is not in the middle, because points in the right side of the plot have higher variance. The remaining two splits are sufficient to model the points on the right reasonably well, but fail to follow the sinusoidal behavior of points in the left part of the plot, in particular, those between $1.5 \simeq x \simeq 2.7$. The second base estimator will try to correct this behavior by assigning a higher weight to points in that interval. The corresponding model succeeds in correcting the predictions in that feature range, but as a result of its limited flexibility, now fails to predict the behavior in the range $0 \simeq x \simeq 1.5$. The third iteration attempts to mediate this behavior, resulting in a rough average between the first

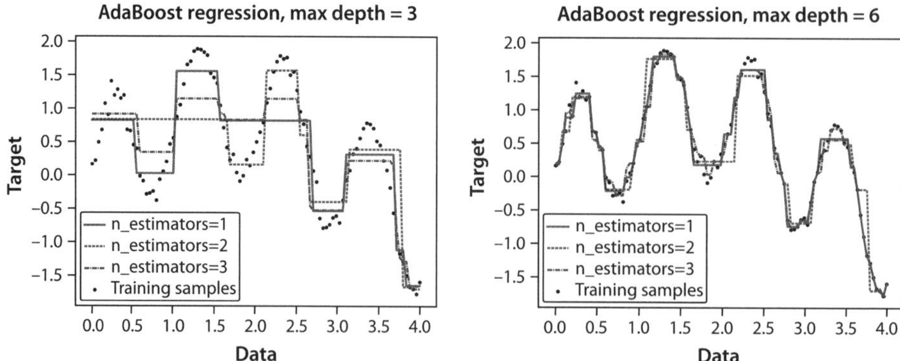


**Figure 6.7: The boosting process in AdaBoost, shown for a decision tree base learner with maximum depth = 3 (left) and maximum depth = 6 (right). Only in the latter case is the base estimator strong enough for the boosting process to be beneficial.**

two attempts that still leaves a lot to be desired in terms of performance, showing that increasing the number of base estimators is not beneficial. You can find code for this section and the next in the lecture notebook "BoostingDecisions.ipynb."

In contrast, if we use a stronger base estimator, a decision tree with maximum depth = 6, the boosting process is beneficial, as shown in the right panel of Figure 6.7. In this case, the base estimator is able to model the behavior well enough that the system of incremental corrections works as planned.

The important lesson we can draw from this simple example is that while the concept of boosting revolves around "weak" learners, *in AdaBoost, there is an accuracy threshold for weak learners below which boosting is not beneficial.* In classification problems, the threshold is usually stated as the learner needs to be better-than-random. In regression, there isn't (to my knowledge) such a clear-cut criterion, but one can empirically check that the base estimators are boost-worthy, as shown in the next section.

However, choosing a more sophisticated base learner, such as a high-depth Decision Tree, might lead to substantial overfitting, which could be alleviated by introducing a regularization term in the loss function. This choice helps avoid placing excessive weight on examples that are hard to predict and increases the robustness of the algorithm. All topics in this section are illustrated in detail in [Meir and Rätsch, 2003].

## 6.5.2 Gradient boosting methods

Gradient boosting methods (or gradient boosting machines, often abbreviated as GBMs) share many elements with AdaBoost: In particular, they still combine several weak learners in an additive fashion, and they work stage-wise (i.e., at each iteration, we don't modify the previous model).

Unlike AdaBoost, however, gradient boosting methods don't assign weights to the examples; instead, *we adjust the target function we are trying to predict.* In the simplest formulation of the algorithm, at each step we fit the residuals (the difference between true and predicted values) of the current model and then add the new model to the current one. Intuitively, the consecutive corrections will fit smaller and smaller residuals, increasing the accuracy of the solution (but often at the risk of overfitting).

Note that fitting the residuals of the current model corresponds to moving along the gradient of the MSE loss function $\sum(y-\hat{y})^2$; its derivative with respect to the predictions $\hat{y}$ is proportional to the residuals, $(y-\hat{y})$. This explains the name of gradient boosting methods and allows us to generalize it for different loss functions, such as the MAE loss or the Huber loss. In each case, the new model we attempt to fit at each step will be the gradient of the loss function evaluated on the points of the training set.

The pseudocode for the original gradient boosting algorithm [Friedman, 2001] is as follows:

1. Initialize the predictions, $\hat{y}_0$, with a constant (typically the mean of the target values, if the loss is the mean square loss);
2. For i = 1, ..., M (number of iterations), do:
3. Compute the negative gradient of the loss function along the training set examples;
4. Fit a new base-learner function $h(x, \theta_t)$ to the negative gradient, by optimizing the parameters $\theta_t$;
5. Update the estimate of the predictor: $f_t = f_{t-1} + \eta h(x, \theta_t)$ ($\eta$ is a parameter that can be chosen in the beginning, as happens in sklearn, or it can be optimized at each iteration).
6. Repeat the procedure of steps 3-5 until some convergence criterion is reached (early stopping), or until the maximum number of iterations is reached.

Figure 6.8 shows the behavior of GBMs on the same simple toy problem used in the previous section. Our weak learner is a tree with maximum depth = 3, and we use the MSE as the loss function. As we keep adding more trees, the combined model is able to match the distribution of target values more and more closely, but it overfits the data very easily. For this reason, it is common to employ some randomization/regularization strategies that can curb this behavior.

#### 6.5.2.1 Minimizing overfitting in gradient boosting methods

Gradient boosting methods are powerful estimators, but they can easily overfit the training data, in particular when decision trees are used as base learners. To mitigate this issue, it is common to use one of the following regularization techniques:

- *Shrinkage*: Choose a small value ($<1$) for the coefficient $\eta$, also known in `sklearn` as *learning rate*, which regulates the contribution of each new correction to the current model. A GBM with smaller values of $\eta$ will take

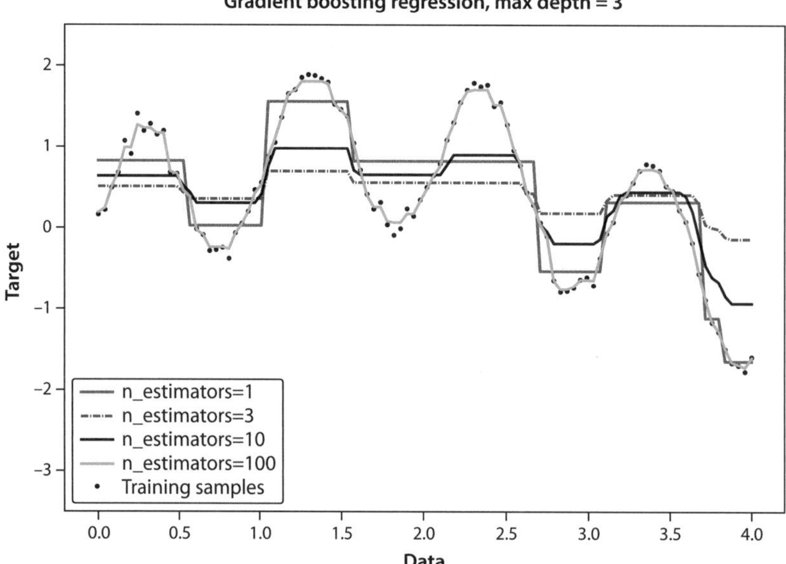


**Figure 6.8: The boosting process in gradient boosting machines, shown for a decision tree base learner with maximum depth = 3 and for the MSE loss function. Unlike in AdaBoost, the process is guaranteed to achieve increasing accuracy as more weak learners are added; however, overfitting is a significant problem that needs to be mitigated.**

longer to achieve the same accuracy; there is a trade-off between the number of iterations and the learning rate.

- *Subsampling*: Select a random fraction of the data (with or without replacement) in order to fit each step. The smaller the fraction is, the stronger the regularization will be.
- *Random feature selection*: Use only a random subset of features at each step. This randomization technique is "borrowed" from bagging algorithms such as Random Forests, where it is similarly used to reduce the variance of the single estimator.

## 6.6 BOOSTING METHODS FOR PHOTOMETRIC REDSHIFTS

We can now make use of boosting methods to tackle the photometric redshifts prediction problem; we will discuss a detailed example using AdaBoost, as well as various methods based on gradient boosting trees.

Several hyperparameters of the algorithm can be explored or optimized. First, we can choose the type of base learner: It is common to use decision trees or stumps, which is what we will show here, but one could use different models, such as local linear models or mini neural networks (see, e.g., [Opitz and Maclin, 1999] and [Bańczyk et al., 2011]).

Second, we can adjust the parameters of the base estimator. In the case of trees, it is customary to choose one of the parameters that limits the complexity of the tree, such as the maximum depth, or the minimum number of examples required in a leaf node or at a splitting point.

Finally, we can tune the *ensembling* parameters: the number of boosting stages or iterations, the learning rate that regulates the contribution of each step to the final model, and the loss function used in the evaluation step.

### 6.6.1 AdaBoost vs GBM

This code snippet shows how to import the algorithm and how to change the parameters in the base estimator, as well as in AdaBoost:

```
from sklearn.ensemble import AdaBoostRegressor
model = AdaBoostRegressor(base_estimator = DecisionTreeRegressor(max_depth = 3),
loss = 'linear', n_estimators = 30, learning_rate = 1)
```

Before optimizing the model, we can verify our empirical intuition that for AdaBoost to be successful, the base learners need to be boost-worthy, and we can also look at the effect of adding more estimators. This is achieved using the "staged predict" property of boosting algorithms in `sklearn`. In code, we can use

```
model.fit(X_train,y_train)
[model.staged_predict(X_test)[i] for i in range(n_estimators)]
```

The top panels of Figure 6.9 show the results of using decision trees of varying depth (with all other parameters kept at default values); it is clear that shallow decision trees are not effective at predicting photometric redshifts, unlike deeper ones. We use the stage-wise prediction to show how the boosting process works and plot the $R^2$ score and the Spearman correlation rank between true and predicted redshifts to assess the performance of the regressors. When decision trees with max depth = 3 are used as base learners, the accuracy of the prediction actually *declines* when adding more stages. When deeper trees (max depth = 6 and 10) are used as base learners, the boosting process is productive and results in improved indicators as more stages are added, until the curves flatten. This behavior, however, is not shared by gradient boosting machines, as noted already in the previous section; even with shallow base learners, the boosting process is beneficial, as shown in the bottom panels of Figure 6.9.

Our last step is to use a grid search technique to find optimal parameters for the AdaBoost regressor. You will find the optimization for the different boosting methods in the notebook "FlavorsOfBoosting.ipynb." Our preliminary exploration suggested that using shallow trees is not beneficial, so we can focus on deeper trees

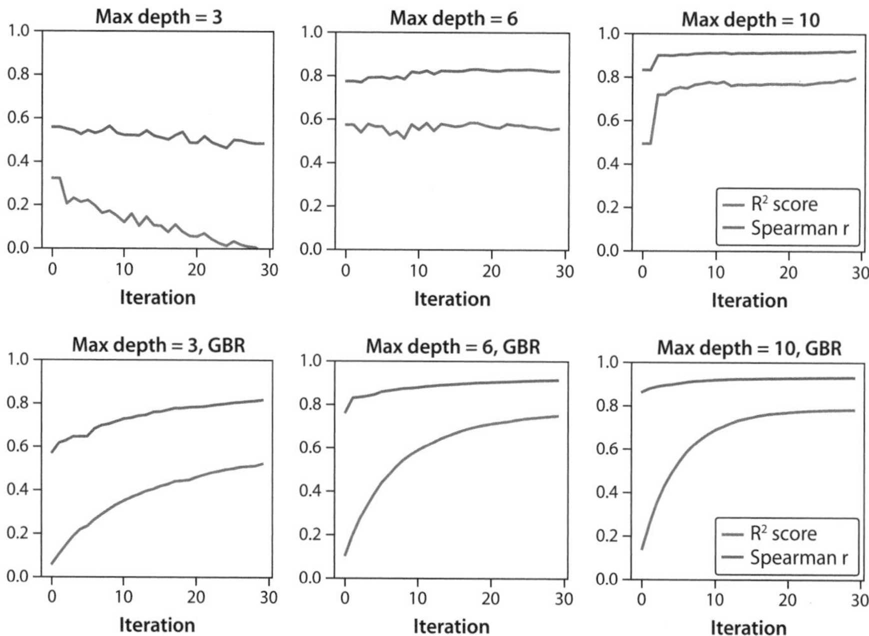


**Figure 6.9: When combined using adaptive boosting (top panels), shallow decision trees (max depth = 3, leftmost panels) are not effective base learners, and the accuracy of the predictions *declines* when adding more estimators, as shown by the declining $R^2$ scores and Spearman correlation factors between true and predicted values. This behavior is not observed in gradient boosting machines (bottom panels), where the boosting process improves results even with weak base learners.**

(max depth $\geq 6$) and vary other hyperparameters, such as the learning rate, the number of estimators, and the loss function:

```
parameters = 'max_depth':[6,10,None],
'loss':['squared_error','absolute_error'], 'n_estimators':[20,50,100],
'learning_rate': [0.1,0.3,0.5]
```

The winning model, shown in the accompanying notebook, consistently uses arbitrarily deep trees; different combinations of loss functions, estimators (50–100), and learning rates give comparable good results.

We can use the best model to generate a few sets of predicted labels (varying the random seed of the CV strategy) to estimate the NMAD and the outlier fraction for the final results, and compare them to those obtained using bagging techniques, such as the Random Forests discussed earlier in this chapter. For the optimal AdaBoost regressor, we obtain (rather consistently) NMAD = $0.032 \pm 0.0003$, and OLF = $0.045 \pm 0.002$, which are actually slightly better than our results using Random Forests.

Our final exercise will be to apply different implementations of gradient boosting methods to this problem and discuss some strategies to speed up the optimization process using a histogram-based version of GBMs, as well as a randomized search of the parameter space.

We begin with `sklearn`'s native implementation of gradient boosting machines; they are available for classification and regression. For our photo-z problem, we will import the latter:

```
from sklearn.ensemble import GradientBoostingRegressor
```

A grid search optimization with the same hyperparameter space used for AdaBoost (reported above) yields similar results, as far as $R^2$ scores are concerned; one important difference is that *Gradient Boosting Regressor (GBR) is about three times as time consuming as AdaBoost was.* This might not seem prohibitive, but two important factors should be kept in mind:

- For gradient boosting methods, using more estimators (in combination with a smaller learning rate) tends to be beneficial, compared to adaptive boosting, which tends to have a plateau in the scores as more rounds are added (as we propose an increasingly difficult problem by focusing on the difficult examples), so we would like to use more estimators.
- There are other parameters, in particular regularization parameters, that should be explored.

We conclude that to fully explore the capabilities of gradient boosting methods, it would be great to speed up the optimization process.

### 6.6.2 Improving optimization efficiency

Our first strategy will be to switch to the histogram-based version of Gradient Boosting Regressor, called (you guessed it) `HistGradientBoosting Regressor` or HistGBR. This algorithm, inspired by Microsoft's LightGBM, works by binning the features into integer-valued bins before building the trees. This trick greatly reduces the number of splitting points to consider and results in a vast reduction of computation time, especially for large data sets. There is, of course, an expected loss in accuracy, because the space of possible splits is reduced; adjusting the binning may help find the optimal trade-off between computational time and accuracy.

We begin by using HistGBR with the default binning, and exactly the same parameters we used for AdaBoost and GBR. Even for this relatively small data set, this is much faster (about 15 times faster than GBR was), suggesting that it is now feasible to explore a wider parameter space. The trade-off is that we obtain a slight decrease in performance compared with GBR, but the standard deviation of test scores over the 5 CV folds suggests that this difference is not statistically significant.

We can define a new hyperparameter space that includes a wider choice of learning rates, more estimators (100/200/500 instead of 20/50/100), and a binary option for early stopping:

```
parameters = 'max_depth':[6,10,None],
'loss':['squared_error','absolute_error'],
'max_iter':[100,200,500], 'learning_rate': [0.05, 0.1,0.3,0.5],
'early_stopping':[True, False]
```

Through this analysis, which is presented in the notebook "FlavorsOfBoosting.ipynb," we can slightly improve on performance by using more trees and a smaller learning rate (as expected). The computational time necessary to explore this much enlarged parameter space is still lower than what was needed for GBR with the more restrictive options, while the performance obtained is similar. Overall, this suggests that HistGBR is much preferable, because exploring a wider parameter space during optimization is a better practice.

A second strategy to improve the efficiency of the optimization process is to use a *random search* of the hyperparameter space, as opposed to a systematic grid search. This choice is motivated by the computational expense of the grid search approach, which makes it impractical to explore all possible combinations, but is also backed up by empirical evidence that the random search approach is likely to identify high-performance regions of the parameter space in a fraction of the time (e.g., [Bergstra et al., 2011; Bergstra and Bengio, 2012]). Therefore, it generally ensures a better (or even just feasible) allocation of resources.

Note that a random search is usually preferable when we have a high-dimensional parameter space; its use is not particularly warranted here. The number of iterations (the number of models that are considered) also needs to be adjusted and depends on the dimensionality of the parameter space as well as the functional dependence of the loss function on the parameters. We will compare the timings with the cell above, where we explore 144 models, and only use 30 for the random search.

In this case, the random search is able to find a comparably good solution in less than 1/5 of the time; as mentioned, the true gains of a random search can be had when exploring high-dimensional spaces and large data sets. It is also possible to use the random search to find the general area of optimal parameters, and then refine the search in that neighborhood using a grid search.

### 6.6.3 XGBoost

Our last boosting algorithm to experiment with is `XGBoost`, which stands for "Extreme Gradient Boosting." It is sometimes known as "regularized" GBM, as it has a default regularization term on the weights of the ensemble, and therefore is considered more robust to overfitting. Additionally, it offers increased flexibility in defining the base learners (which are not limited to decision trees), the regularization strategy, and the objective (loss) function (note that this doesn't apply to the base estimators, e.g., how splits in trees are chosen, but refers to the loss used to compute pseudo-residuals and gradients).

In the notebook "FlavorsOfBoosting.ipynb," we show that XGBoost is slightly more efficient than GBR and achieves comparable results on a similar grid. The same behavior (similar performance and slightly increased efficiency with respect to GBR) is observed when we use random search to explore additional models with more trees, a wider choice of learning rates, and add subsampling as an extra form of regularization.

We conclude that boosting algorithms are a competitive technique for this task, and all the boosting algorithms we introduced offer similar performance; the best choice in this case may be to simply use the fastest method (in this case, HistGBR, possibly in combination with a random search of the hyperparameter space).

## 6.7 FEATURE IMPORTANCE

One of the most useful features (pun not intended) of tree-based methods, including bagging and boosting methods where the base learner is a tree, is that they are highly interpretable: We can actually visualize exactly the criteria that they use when making decisions. Because of this, we can also gather information about which features have the highest impact on a model's decision. The degree to which each feature affects the final outcome is known as *feature importance*. Obtaining a quantitative assessment of the importance, however, is not trivial, and the result is at least in part algorithm dependent.

In tree-based methods, we can intuitively see that the most important features will come "early" in the hierarchy of the tree, because discriminating along important features leads to large decreases of impurity on a large fraction of the data. To describe this mathematically, we can look up all the splits associated with a certain variable, sum up the decrease of impurity induced at each split, and thus obtain a numerical estimate of that feature's importance. We can do this exercise for decision trees, of course, but because they are prone to overfitting, it might be better to use a more robust model, such as the ensemble models described in this chapter. All tree-based methods in `sklearn`, once fitted, have an attribute called `feature_importance` that contains a normalized version of the feature importance described above. Here is a fun example[5] that shows a feature ranking of the different pixels in images used for facial recognition; they roughly agree with our intuition that features like eyes, mouth, or edges would be most important.

5 https://scikit-learn.org/stable/auto_examples/ensemble/plot_forest_importances_faces.html

Feature importance is useful for two main reasons. The first one is to create some intuition about which features are most relevant. If our data come from experiments, as it is common in physics or astronomy, understanding which observations are most useful to make in order to obtain a highly predictive model is very important. The second reason is that a ranking of the features in order of importance could be used to get rid of less relevant ones without a substantial loss of information. This could help reduce the variance caused by noisy features, or as a dimensionality reduction technique, as noted again in Chapter 7.

Feature importance and rankings calculated through the mean decrease of impurity (i.e., as described earlier in this section) come with a few caveats. The first one is that high cardinality features, having many unique values, tend to be picked more often, and their importance can be overestimated; see an example here.[6] The second issue is that robustness is quite important: If we are considering a model that overfits, we shouldn't trust the ranking, because we know that the model is picking up a lot of noise. The last one, which is actually the worst issue in my experience as a practitioner, is that highly correlated features will "split" the importance. If you imagine having two features that are copies of one another, each of the decision trees will choose to split along either of them with equal probability, resulting in an importance that is equally divided between the two and thus lower than expected.

A possible solution for the first caveat is to use *permutation-based* feature importance, which is the decrease in a model score when a single feature value is randomly shuffled. This is not helpful, however, for the third issue. The problem of assessing importance for highly correlated features can only be solved, in my opinion, by reducing the correlation among features. One possibility is to use a clustering algorithm to assess features that are similar, and then just choose one feature per cluster as the representative of each group (e.g., [Chormunge and Jena, 2018]). Finally, the second caveat can be resolved by deriving the feature importance not necessarily for the "best" model (the model with the highest test scores), but for a robust (low-bias) model. Note that all the methods of assessing feature importance are model dependent; if we switch algorithms, we can't expect the feature ranking to stay the same, so it may be useful to look at feature importance for different methods and average them if we are interested in a model-agnostic assessment.

An alternative possibility to feature ranking using an importance measure is SHapley Additive exPlanations, or SHAP, values [Lundberg and Lee, 2017]. The SHAP values attempt to quantify the contribution of each feature ("player," in game theory parlance) to the final decision of the model. Unlike the impurity-based feature importance, SHAP values can be calculated for any algorithm; recent work shows that for tree-based methods, the SHAP values are equivalent to the impurity-based feature importance when the parameter `max_features` is set to 1 [Sutera

6 https://scikit-learn.org/stable/auto_examples/inspection/plot_permutation_importance.html#sphx-glr-auto-examples-inspection-plot-permutation-importance-py

et al., 2021]. The Python package `shap` is available here[7] and can be integrated with all ML models.

### 6.7.1 Uncertainty estimation

One very important (and highly nontrivial) piece of providing predictions in a supervised learning model is to estimate the uncertainty associated with individual predictions. The evaluation metrics that we use, from MSE to complex loss functions, are only *aggregate* reliability scores, while we'd like to assess the reliability of our predictions for every instance (in this case, for every galaxy).

We can start by thinking about the possible sources of uncertainty. Some of them are associated with noise in the data: for example, the experimental uncertainty in each broadband flux measurement used as input. We can imagine that even if the model were exactly the same, the estimates of photometric redshifts obtained from very noisy measurements from a ground-based telescope should reflect a higher uncertainty than those obtained from higher SNR measurements obtained from a space-based telescope. Another possible cause of uncertainty is the fact that we have limited data: in our case, only six broad-band measurements. The finite size of our learning set is another limitation. There are also uncertainties associated with the model itself, for example, the fact that we have chosen a certain model (e.g., a boosting model) with certain hyperparameters that have uncertainties associated with them.

In the ML community, it is common to characterize the uncertainties as either *aleatory* or *epistemic*. We characterize a source of uncertainty as epistemic if it's possible to reduce it by collecting more data or improving our ML model (e.g., by switching magnitude with colors, as discussed in one of the exercises). We characterize a source of uncertainly as aleatory if it is irreducible, that is, collecting more data or changing our model would not help. For example, experimental noise is typically associated with aleatory uncertainty. Epistemic and aleatory uncertainty tend to align with the concepts of *systematic* and *statistical* uncertainty, respectively, used in the Physics community; see this fun Twitter thread.[8]

Importantly though, the distinction between the two classes is not obvious, as noted, for example, in [Der Kiureghian and Ditlevsen, 2009]. For example, we may consider degeneracy as a source of aleatory uncertainty: Given the measurements we have, there may be two galaxies that have the same photometric luminosity within our errors but different redshifts. Collecting more data for other galaxies would not help, reinforcing our idea that this is an aleatory uncertainty. But we could also collect more data as additional photometry in other bands, which could break the degeneracy and lead to improvement. In this case, this type of uncertainty

---

7 https://github.com/slundberg/shap

8 https://twitter.com/zacharylipton/status/1277396285914525696

would be better characterized as epistemic. This point is also made very nicely in the recent review [Hüllermeier and Waegeman, 2021], where the authors note that in general, "embedding data in a higher-dimensional space will reduce aleatoric and increase epistemic uncertainty, because fitting a model will become more difficult and require more data." The bottom line is that the distinction is blurry, but it may still be useful to think about sources of uncertainty as reducible vs irreducible for a given problem, to understand what steps we can take to improve predictive uncertainties.

Going back to our problem, though, how can we estimate the total uncertainty in our photometric redshifts on an instance-by-instance basis?

For the aleatoric part, one possible approach is to assume a noise model for each of our observed inputs; for example, in a regime of decent SNR, a Gaussian distribution where the mean is equal to the observed value, and the standard deviation is equal to the experimental error on each measurement. We can then use Monte Carlo sampling from the noisy input distributions to obtain distributions of our predicted output and treat them as approximate probability distribution when deriving estimated confidence intervals.

The uncertainty estimate obtained with this method will only reflect part of the model-based or epistemic uncertainty; for example, it does not include any uncertainty associated with the choice of parameters in the model. To estimate this part of the uncertainty, we need to train a model to output not just a point-based estimate but also a probability distribution of possible outputs that reflects the model uncertainty. A lot of the literature on uncertainty estimation targets the model-based or epistemic uncertainty in deep neural networks (see, e.g., [Gawlikowski et al., 2021]); examples of those methods include Bayesian neural networks (e.g., [Mullachery et al., 2018]), and ensembling of deep neural networks (e.g., [Lakshminarayanan et al., 2016]). This last idea of using ensembles of predictors is the basis of the recently developed algorithm `NGBoost` [Duan et al., 2020], which proposes a method for generating probabilistic estimates of output even when the base predictor is not a neural network, but a simpler regressor, such as a decision tree or a linear model. The distributions generated by `NGBoost` are instance-specific, for example, because objects in some areas of feature space are better understood and represented in the training domain than others. Using `NGBoost` could be a way to calculate the epistemic uncertainty to be added in quadrature to the aleatory uncertainty obtained via Monte Carlo sampling of noisy inputs.

## 6.8 LESSONS LEARNED

This chapter introduced the notion of *ensembling* as a way of increasing accuracy and robustness of ML models. We have used decision trees as our base learners because of their versatility, but other base learners, from linear regressors to neural networks, can be used as well. Here is a summary of our main findings:

- We have considered bagging methods, such as Random Forests and Extremely Random Trees, and boosting methods like AdaBoost, GBRs, and XGBoost. The former are based on averaging independent, fully developed trees; the latter are sequential methods that seek to refine predictions iteratively.
- In Random Forests or Extremely Random Trees, the goal is to decrease variance through the averaging process. This implies an increase in bias because of the randomization process, but we can hope that the total balance will decrease the total error. As a result, bagging methods are typically used to reduce overfitting.
- In Random Forests, the training set is randomized by means of sampling with replacement, and the subset of features considered at any split is also randomized.
- In Extremely Random Trees, the subset of features considered at any split is randomized, and the threshold used in the splits is also chosen at random.
- In boosting methods, the improvement is obtained by modifying the training set (AdaBoost) or the loss function (gradient boosted methods, such as GBR or XGBoost).
- In AdaBoost, the system focuses on solving difficult examples, that is, the ones that currently have the highest errors. These samples participate with increased probability to the following iteration of the training set. The problem becomes increasingly difficult to solve, so convergence is obtained when the normalized loss function reaches a *maximum* value, unlike in many iteratively built models that are based on minimizing the loss.
- In gradient boosting methods (GBR, XGBoost, and the like), the loss function is modified in each round, with the goal of fitting the residuals of the previous model. Because of this "local loss" approach, they can improve accuracy very fast even when the base learners are weak, but they are prone to overfitting. This can be solved by appropriate hyperparameter optimization; overall, they are among the most powerful ML algorithms.
- When the exploration of the hyperparameter space is prohibitive from a computational resource perspective, it is useful to use the "discretized" version of GBR, named HistGBR, where the features are binned before processing, greatly reducing the number of possible splits in the base decision trees, and to use a random search of the hyperparameter space, exploring a fraction of the possible models instead of conducting a systematic grid search.

To learn more about bagging and boosting methods, I'll recommend again [Louppe, 2014], and [Meir and Rätsch, 2003] and [Natekin and Knoll, 2013], respectively.

Finally, this chapter introduced two important concepts that go beyond the algorithms presented: (1) estimating uncertainties on the predictions made by our models, and (2) ranking the features of our models in order to understand how decisions are made and as a means of reducing the dimensionality of the data using feature selection.

## 6.9 REVIEW AND DISCUSSION QUESTIONS

Note: Questions and exercises marked by ** are more complex, open-ended, or time consuming. Those marked by * have more than one correct answer.

**Exercise 1.** Reflect on the differences between RF and ERT. In a situation of high variance, which one would you expect to be most helpful? Why?

**Exercise 2.** What is the main advantage of using Random Forests in place of single decision trees?

A) Lower the variance
B) Lower the bias
C) Shorten computing time
D) Obtain more interpretable results

**Exercise 3.** Attribute each of the following statements to bagging methods or boosting methods:

A) They tend to lower the variance.
B) They tend to lower the bias.
C) The final estimator predicts a simple average of all the components.
D) The final estimator predicts a weighted average of all the components.
E) Base estimators are independent of each other.
F) Base estimators are built on the basis of previous stages, results.

**Exercise 4.** * What is randomized in Random Forests?

A) The data, via bootstrap of samples
B) The features, since we select a random set of them to make splits on
C) The threshold at which a split is made
D) The trees that participate in the final ensemble classifier/regressor

**Exercise 5.** Attribute each statement to GBMs, AdaBoost, or both:

A) The training set changes at every iteration.
B) The target is adjusted at every iteration.
C) Gradient descent can be used to find the minimum of the loss function.
D) There is a threshold in boost-worthiness of the base learner.

**Exercise 6.** * What is randomized in Extra Random Trees?

A) The data, via bootstrap of samples
B) The features, since we select a random set of them to make splits on
C) The thresholds at which a split is made
D) The trees that participate in the final ensemble classifier/regressor

**Exercise 7.** How is the feature importance calculated in tree-based methods?

A) Reporting the correlation coefficient between each feature and the target in the final ensemble
B) Calculating the mean decrease of impurity across splits that use that feature
C) Calculating the average of the correlation coefficient between each feature and the target in a random selection of trees
D) Listing the features in alphabetical order

**Exercise 8.** How is the global performance of a Random Forest/Extra Random Tree method assessed?

A) Using the best performing tree in the ensemble

B) Averaging the performance of the 50% best-performing trees
C) Reporting the performance of the worst-performing tree, to be cautious
D) Averaging the performance of all trees

**Exercise 9.** * Which of these parameters, if decreased, can help reduce the gap between training and test scores in a bagging ensemble method?

A) Max_depth
B) N_estimators
C) Min_sample_splits
D) Min_sample_leaf

**Exercise 10.** Which of these parameters, if increased, will definitely improve (or at least not reduce) the performance of a bagging ensemble method?

A) Max_depth
B) N_estimators
C) Min_sample_splits
D) Min_sample_leaf

**Exercise 11.** Which of these estimators is most likely to work well even with a very weak base learner?

A) Random Forests
B) Extremely Random Trees
C) AdaBoost
D) GBMs

## 6.10 PROGRAMMING EXERCISES

**Exercise 1.** Start from the full data set of photometry vs redshift in the file "DEEP2_uniq_Terapix_Subaru_v1.fits." We saw in the lecture notebook "Photoz_RandomForest.ipynb" that the performance changes a lot once the selection criteria are applied. Figure out (by looking at it step by step) which of the data cleaning cuts we made was the most significant in terms of improving the scores of the final model.

**Exercise 2.** Using the data set selection implemented in the notebook "Photoz_RandomForest.ipynb",

1. Optimize (using a grid search for the parameters you deem to be most relevant) the Extremely Random Trees algorithm and compute the outlier fraction and NMAD. How does it compare to the optimal Random Forests model? Comment not just on the scoring parameter(s), but also on variance/bias. Which one would you pick?
2. ** In the paper used as a reference [Zhou et al., 2019], the authors use colors, not magnitudes, as features (or to be precise: one magnitude and five colors). Find in the paper the exact list of features, and generate the new features to match what is done there.
3. ** Armed with your new set of features, use a bagging or boosting algorithm of your choice to match and possibly beat the performance quoted in the paper (NMAD: 0.0174; OLF:4.54%).

**Exercise 3.** Implement AdaBoost for the photometric redshifts problem with a different type of base learner (e.g., a linear regressor or a polynomial regressor). What changes? Which one would you recommend using?

CHAPTER SEVEN

# 7 Clustering and Dimensionality Reduction

*There was once a girl from West Reading*
*Whose data were too big for keeping*
*So to keep them at bay*
*She did PCA*
*But they lost all their physical meaning.*

In this chapter, we take a quick dive into the realm of unsupervised learning. Unsupervised learning is all about exploration: We can mine the data for patterns, find similarities and differences, and discover relationships; we can also look for lower-dimensional representation of the data, to decrease computational burden or to facilitate visualization of complex dynamics. Clustering and dimensionality reduction techniques are very flexible tools; often we will want to use them in conjunction with supervised techniques to simplify a supervised learning problem before solving it, or simply for the preliminary data exploration that, as we have seen, is a fundamental step in any ML problem-solving pipeline.

## 7.1 CLUSTERING

Clustering is the branch of unsupervised learning that is concerned with organizing the data in groups whose members share some measure of similarity.

We see examples of clustering in everyday life. For example, when browsing the news, articles with similar topics are often grouped together; applications on our phones can be automatically organized in folders, based on some similarity in content (financial, health-related, utilities, etc.). These groupings happen without specifying the categories (i.e., the labels); groups are formed by exploring the feature space, for example, words or sentences in the case of news articles, and noting areas where the density of samples is higher, which signals that several samples have some attributes (features) in common. Another example, which we have sadly

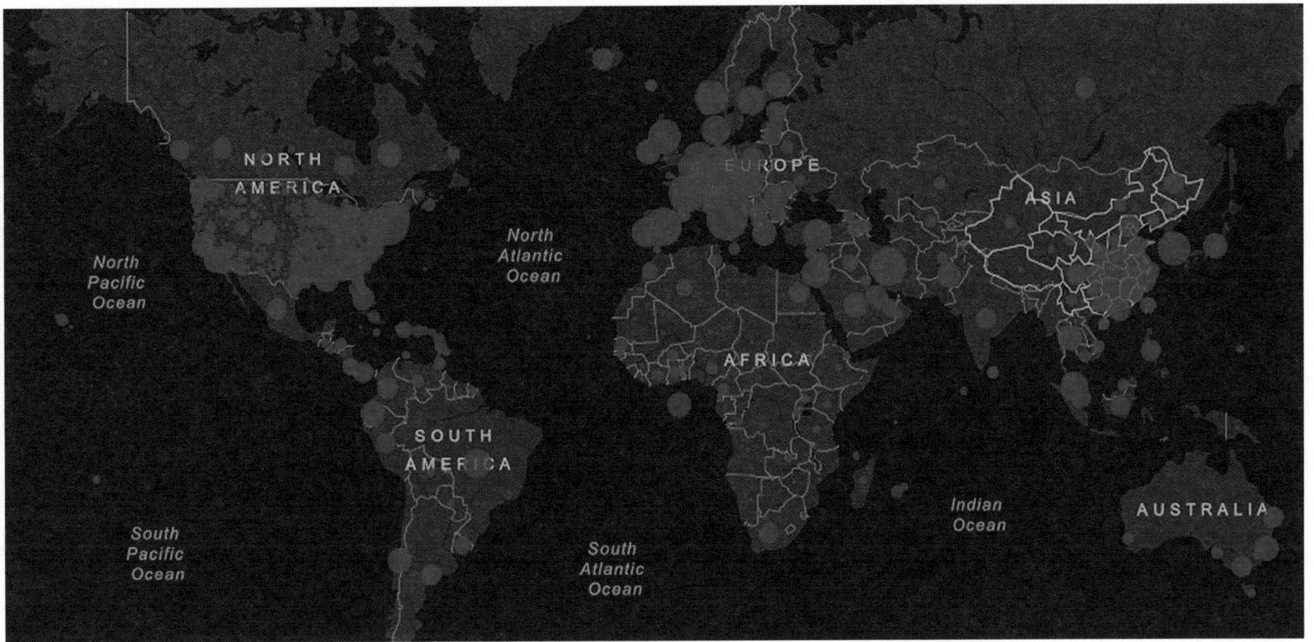


**Figure 7.1: A clustered visualization of the SARS-COV2 virus spread in 2019. The cluster centers trace areas of high density of infection,and their radii are proportional to the size of each cluster.**

become very familiar with in the past few years, is the visualization of outbreaks in a specific disease (see, e.g., [Megahed et al., 2020]). Figure 7.1 shows a clustered visualization of the SARS-COV2 virus spread in 2019. The cluster centers trace areas of high density of infection, and their radii are proportional to the size (number of members) of each cluster. Finally, a common application of clustering for the physical sciences is *anomaly detection*: We learn what are the typical categories that data belong to, so that we can quickly and effectively recognize outliers or unusual events in a data stream (see, e.g., the bibliography in [Baron, 2019] and [Nachman, 2020]).

### 7.1.1 Objective

In general, the goal of a clustering algorithm is to highlight similarities and differences in the data.

Clustering can be formulated as an *optimization* problem: We can choose an objective or cost function, similar to the loss functions of supervised learning methods, which measures the quality of a clustering structure. Then we try to find the partition that maximizes the objective function or, equivalently, minimizes the cost function. The cost function typically depends on a measure of similarity in feature space. The most common choice is the Euclidean distance between points, but others measures of distance are also valid.

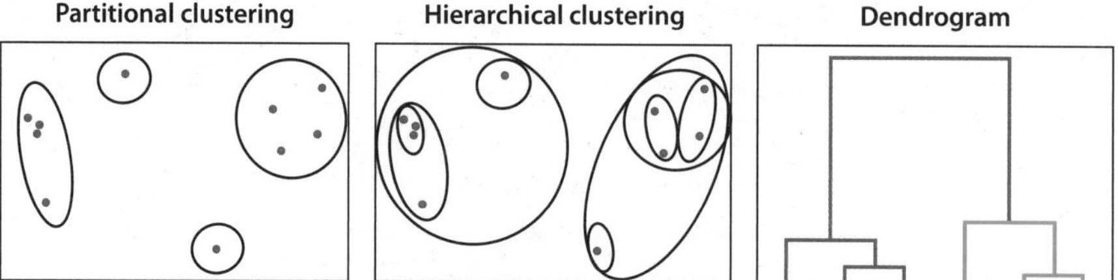


**Figure 7.2: In partitional clustering (left), each point belongs to one and only one cluster, unlike in hierarchical clustering (middle), where nested divisions are created at each step. The structure created by hierarchical clustering methods is often shown as a *dendrogram* (right).**

There are several possible scaffoldings that can be used to organize a clustering structure. One common distinction is between *partitional* clustering, where each object belongs to one and only one cluster, and *hierarchical* clustering, where clusters can be "nested" within one another. The diagram that shows the organization of a hierarchical clustering in groups is called a *dendrogram*. The difference between the two types of clustering is illustrated in Figure 7.2.

Partitional and hierarchical clustering methods have different cost functions; in particular, the former will have a "global" cost function, which is calculated on the entire data set, while the latter have a "local" cost function, since subsequent clusters are created on a subdivision of the data set. We have already seen this trait in supervised learning methods: Some of them have global loss functions (e.g., linear models), while others have local ones (e.g., tree-based methods, where the criterion to evaluate the quality of a split is calculated on the current node).

In either case, once the clustering algorithm has been set up as an optimization problem, how can we go about solving it? If our data set is finite, the possible clustering solutions (i.e., the possible ways to organize all the samples in clusters) are also finite, so it would be tempting to just enumerate all of them and choose the one that corresponds to the smallest cost function. However, this problem is NP-hard (e.g., see [Van Leeuwen, 1991]), and therefore its computational complexity is prohibitive even for moderately sized data sets. The following sections introduce some popular clustering algorithms and show how they explore the solution space in intelligent ways to improve on the complexity of the "naive" solution.

### 7.1.2 *k*-means

One of the simplest, yet powerful algorithms for clustering is the $k$-means algorithm. It is a partitional algorithm, where each point in the data set is assigned to one of $k$ clusters; the number of expected clusters $k$ needs to be specified beforehand.

For a given partition in clusters, a *centroid* can be identified for each cluster. Each cluster's centroid is found by averaging the coordinates of all the points that are currently assigned to that cluster (this is where the "means" of $k$-means comes from). The objective of the clustering algorithm is to minimize the sum of squared distances of each point from its assigned centroid. If the distance metric is the

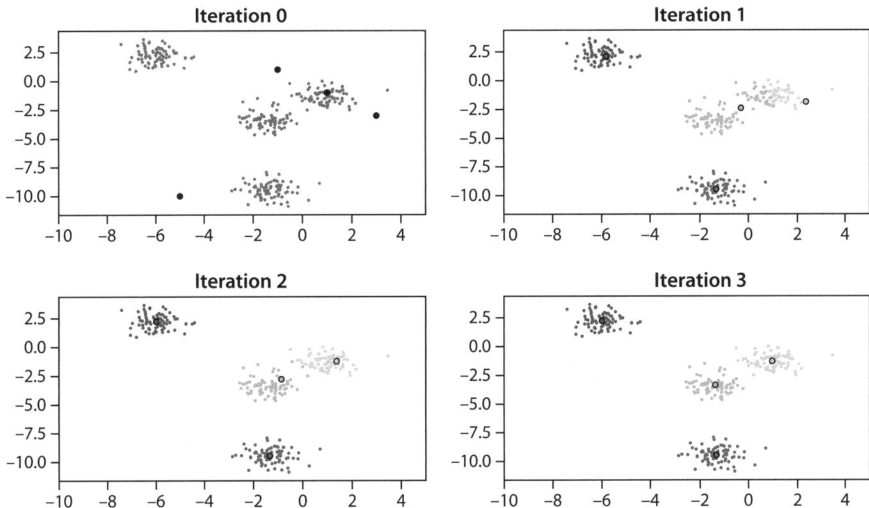


**Figure 7.3: The first four iterations for a *k*-means clustering with four easily recognizable clusters (and $k = 4$). Iteration 0 shows the initial centroids. At each iteration, points are assigned to the closest current centroid, and the new centroids are calculated and shown as larger points with a black contour. The colors show the current membership of each point.**

Euclidean distance, the cost function $J$ becomes:

$$J = \sum_{i=1}^{k} \sum_{x \in C_i} (x - C_i)^2, \tag{7.1}$$

where $C_i$ is the vector containing the coordinates of the centroid of cluster $i$. This cost function is also called *inertia* in `sklearn`.

The pseudocode for the $k$-means algorithm is an example of an expectation-maximization process and is very simple:

- First, select $k$ data points as the initial guess for the $k$ centroids.
- Then build a clustering partition by assigning each data point to the closest centroid.
- Next, recalculate the coordinates of the $k$ centroids (by averaging the coordinates of all the points currently assigned to each cluster).
- Repeat the last two steps until the membership of the clusters does not change (or some other convergence criterion is met).

It is worth noting that while centroids can be thought of as a typical or representative member of each cluster, they won't in general coincide with any element of the data set.

Figure 7.3 shows the first four iterations for a $k$-means clustering with four easily recognizable clusters (and $k = 4$). For clarity, we select the original centroids

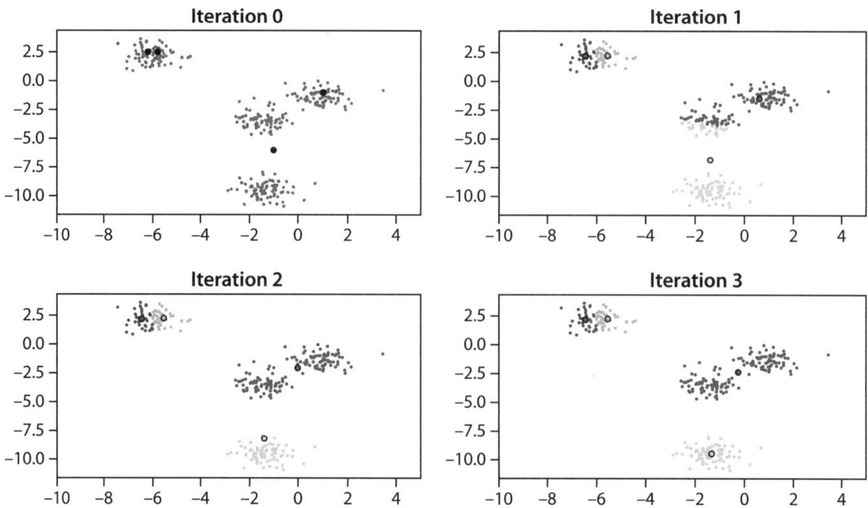


**Figure 7.4: The first four iterations for a $k$-means clustering for the same clusters as in Figure 7.3, but for a poor choice of initial centroids. The algorithm does not converge to the correct solution.**

manually. At each iteration, points are assigned to the closest current centroid, and the new centroids are calculated and shown as larger points with a black contour. The colors show the current membership of each point. Iteration 0 shows the initial centroids. In iteration 1, all the points in the top-left cluster are assigned to the centroid with coordinates $(-1, 0)$, and all the points in the bottom cluster are assigned to the centroid with coordinates $(-5, -1)$; these two become stable clusters immediately, with centroids roughly in the middle of each cluster of points. The two middle clusters require a few more iterations to acquire stable membership; at iteration 1, the centroids are still offset from the centers of the two clusters, and as a result, the membership at iteration 2 is still incorrect, as some points from the rightmost cluster are assigned to the centroid to their left. By iteration 3, however, the algorithm has correctly identified all centroids, and the cluster membership has become stable.

This example is a fairly simple one: Clusters are well separated from one another, and there is no ambiguity in identifying them. $k$-means was able to identify those clusters easily, and just in a few iterations; in fact, the complexity of $k$-means is linear in all relevant variables, $\mathcal{O}(n \times k \times I \times m)$, which makes it a very fast algorithm even for large data sets. Here $n$ is the total number of points, $k$ is the number of clusters, $I$ is the number of iterations, and $m$ is the dimensionality of feature space.

However, $k$-means also has several limitations. An important one is that even for simple cases such as this one, it is sensitive to the initial conditions for centroids, and it might not converge to the correct solution if the centroids are not well separated. Figure 7.4 shows the $k$-means solution for the same data set but for a different initial choice of centroids. The presence of the two close centroids in the vicinity of the top-left clusters throws off the cluster assignment, and because the total number of clusters is constant, membership of the other clusters and centroid placement are also negatively affected. The algorithm does not converge to the correct solution.

The good news is that this particular problem can be overcome relatively easily. One strategy is to select the initial centroids not at random but by spacing the centroids as far apart as possible (in other words, the first centroid is one data point selected randomly, the second centroid is the point in the data set that is farthest from the first, and so on). This is sometimes referred to as the *k*-means++ algorithm. Another strategy is to run the algorithm several times with different random centroid initializations and choose the solution with the smallest inertia. In fact, the default settings in `sklearn` include both these options. However, there are still many situations where *k*-means cannot perform well. We discuss some of them in the following sections.

### 7.1.3 The limitations of *k*-means

The main limitation of *k*-means is that the algorithm can only form clusters of globular shape. This shortcoming can be understood easily by looking at its cost function, Eq. 7.1: the distances in all dimensions are weighed equally, and the contribution of each cluster is the same, independent of membership size or compactness. As a result, *distributions that are elongated or nonconvex are not correctly recognized by k-means*. For example, in the smiley face shown in the left panel of Figure 7.5, it is easy for a human to pick out the four clusters. However, *k*-means is unable to assign centroids and determine cluster membership correctly, even when the correct number of clusters is specified initially, because it will simply divide the space into four convex shapes of similar size. Additionally, even globular-shaped clusters of different sizes and densities might present issues: When clusters overlap, each point is simply assigned to the closest centroid, as shown in the right panel of Figure 7.5, where we see that many points from the central large cluster are assigned to the smaller side clusters. Although this is not entirely avoidable, as the clusters overlap in feature

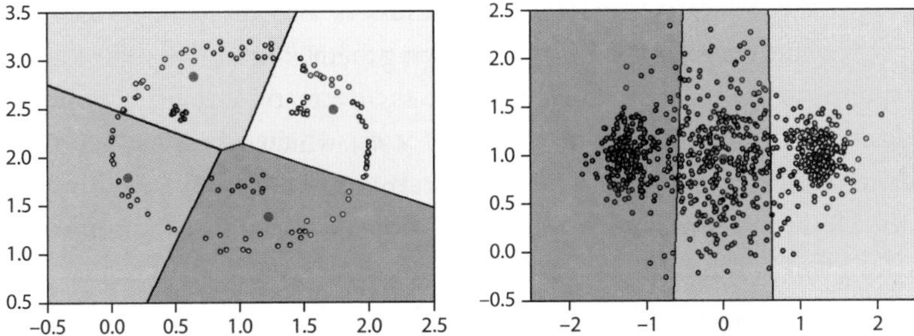


**Figure 7.5: (Left) *k*-means is unable to assign centroids and determine cluster membership correctly and will divide the space in *k* convex regions of similar size. (Right) There is no "perfect" solution for overlapping clusters, but *k*-means is unable to retain the notion that many assignments are highly uncertain.**

space, it would be better to retain some notion of this uncertainty in our clustering scheme, for example, by means of a probabilistic membership assignment.

Finally, a limitation of $k$-means and many other clustering algorithms is that *the number of clusters needs to be specified beforehand.* When we don't have a good sense of how many categories of objects are present in the data, this limitation is not ideal. A possible solution could be to treat the number of clusters $k$ as a hyperparameter, build a clustering scheme for several values of $k$, and choose the one that gives the best result. However, the challenge here is that unlike the case of supervised learning methods, where evaluation metrics can easily be created by comparing ground truth labels and model predictions, evaluating the quality of a clustering scheme is more complex. In most supervised model evaluation frameworks, we can pick the hyperparameter values that correspond to the lowest error or highest score. We can't do the same thing for $k$-means: As we increase the number of clusters, the cost function will always decrease, because each point would be assigned to a centroid that is closer. In the limiting case of as many clusters as data points, the cost function is exactly zero (as each point is its own cluster centroid), but we would probably all agree that such a clustering scheme is quite useless. What can we do? One possible approach is to formulate a different evaluation function for this problem, which takes into account not just the inertia of each clustering structure but also its complexity. We discuss two possible choices in the next section, and a third one later in Section 7.3.2.

### 7.1.4 Evaluating the quality of clustering algorithms: The elbow method and the silhouette score

I mentioned earlier that often we don't know a priori how many clusters to expect in the data. One simple indication of an adequate number comes from the so-called *elbow method*. The idea behind it is to build a clustering scheme for several values of $k$ and plot the inertia (the cost function of Equation 7.1) as a function of $k$. The curve will be monotonically decreasing, as mentioned, because a higher number of clusters will lead to smaller distances between data points and their assigned centroids. However, we can try to identify the $k$ that corresponds to the *largest change in the slope* of the curve (formally, an inflection point in the second derivative). This corresponds to the configuration for which the decrease of the cost function *per new cluster* is maximized. Note that in many cases, (1) the elbow method provides a ballpark indication, rather than a definite answer, and (2) it implicitly assumes that the clustering solution provided by $k$-means makes sense. For example, for very elongated or nonconvex distributions, we know that $k$-means cannot provide a sensible clustering solution, and as a result, the elbow method won't provide a sensible answer. This case is particularly tricky because in high-dimensional data, we can't visualize clusters beforehand, which makes it hard to know whether $k$-means is an appropriate clustering algorithm. Figure 7.6 shows the elbow curves for the two distributions shown in Figure 7.5. We can make a few observations. First,

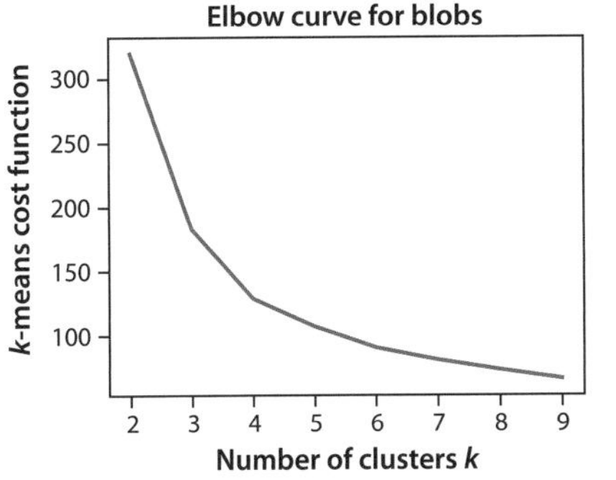


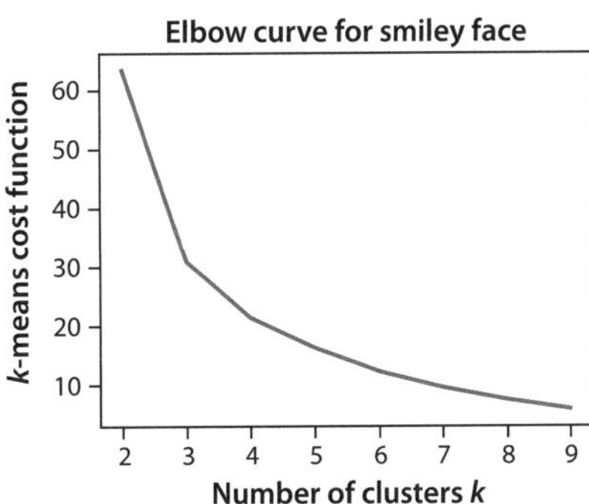


**Figure 7.6: Elbow curves for the two distributions shown in Figure 7.5. Note that the definition of "best *k*" is often somewhat ambiguous. For both these cases, $k = 3$ and $k = 4$ appear to be good candidates.**

because this is a discrete curve, singling out the suggested value of $k$ isn't always straightforward: in both cases, it could be 3 or 4. Squinting and trusting our by-eye-curve-fitting-and-differentiating skills, we could claim that for both curves, the suggested value is 3. Unsurprisingly, this is correct in the case of the blobs, which are more $k$-means friendly, at least in terms of shape, and incorrect in the case of the smiley face, where $k$-means can't isolate the four distinct distributions and instead "splits" the points from the outer circle among the other three sets of points.

One problematic aspect of the elbow method is that it gives an idea of the best number of clusters *within the limits of a certain method,* but it does not evaluate the overall quality of the clustering scheme that we have created. Are the clusters well separated? Are points in each cluster close to their centroid? We can't use the elbow method to answer these questions, but we can use a different quality measure, known as the *silhouette score.*

This indicator aims to measure how reliable the assignment of each object to its cluster is by comparing the mean intra-cluster distance to the mean distance to the nearest cluster. For each data point $i$ in cluster $C_i$,

$$a(i) = \frac{1}{N(C_i) - 1} \sum_{j \in C_i, i \neq j} d_{ij} \tag{7.2}$$

is the mean distance between that point and the other members of the same clusters; $N(C_i)$ is the number of objects assigned to cluster $C_i$. We can interpret $a(i)$ as a measure of how decisively the data point $i$ is assigned to its cluster: the smaller the value, the stronger the assignment will be.

Similarly, we can calculate, for the same data point $i$, the mean distance to the nearest (distinct) cluster:

$$b(i) = min_{k \neq i} \frac{1}{N(C_k)} \sum_{j \in C_k} d_{ij}. \tag{7.3}$$

We are finally ready to calculate the silhouette score as

$$s(i) = \frac{b(i) - a(i)}{max\{a(i), b(i)\}}. \tag{7.4}$$

The silhouette score is bound between $-1$ and 1. When $a(i) \ll b(i)$, the point is much closer to other points in the same cluster than it is to points in the next nearest cluster. This indicates that the clustering works well, and in this case, the silhouette score value will be $s(i) \sim 1$, because both numerator and denominator will be $\sim b(i)$. In contrast, $b(i) \ll a(i)$ indicates a poor choice of cluster assignment, because the data point is closer to objects in a different cluster, compared to its own. The silhouette score for such points will be $s(i) \sim -1$, as the numerator will be $\sim -a(i)$ and the denominator will be $\sim b(i)$. Finally, when $a(i) \sim b(i)$, the assignment is dubious, indicating that clusters are overlapping; this corresponds to $s(i) \sim 0$. By plotting the silhouette score distribution for all points in a data set, we can have a snapshot of how well the clustering scheme is working. We can also calculate the average silhouette score for the entire data set as a summary statistic, use it to compare the quality of different clustering schemes, and select an optimal number of clusters, similarly to what we did for the elbow method.

## 7.2 DENSITY-BASED CLUSTERING

We have learned that one of the limitations of *k*-means is that it can only detect clusters of convex (and in fact, globular) shapes. A different category of clustering algorithms that can detect clusters of arbitrary shapes is *Density-Based* Clustering; the most popular is probably DBSCAN, which stands for Density-Based Spatial Clustering of Applications with Noise [Ester et al., 1996].

DBSCAN works by identifying clusters as areas of high density, separated by regions of lower density. Just as for *k*-means, we will need to select a distance measure; however, we won't be required to select the number of clusters beforehand. DBSCAN is characterized by two parameters: a threshold for proximity, often called $\epsilon$, and the minimum number of points in a high-density neighborhood, $n_{\min}$. The algorithm will assign to each point one of three labels. A *core* sample has at least $n_{\min}$ other samples within a distance of $\epsilon$: core samples live in high-density areas. A "border" sample is a neighbor (i.e., is within distance $\epsilon$) of a core sample, but is not itself a core sample. Finally, "noise" points are not within a distance $\epsilon$ of any other point, and they can be interpreted as outliers.

The pseudocode for DBSCAN (inspired by these lecture notes[1]) is as follows:

---

1 https://www.cs.ubc.ca/~schmidtm/Courses/340-F15/L9.pdf

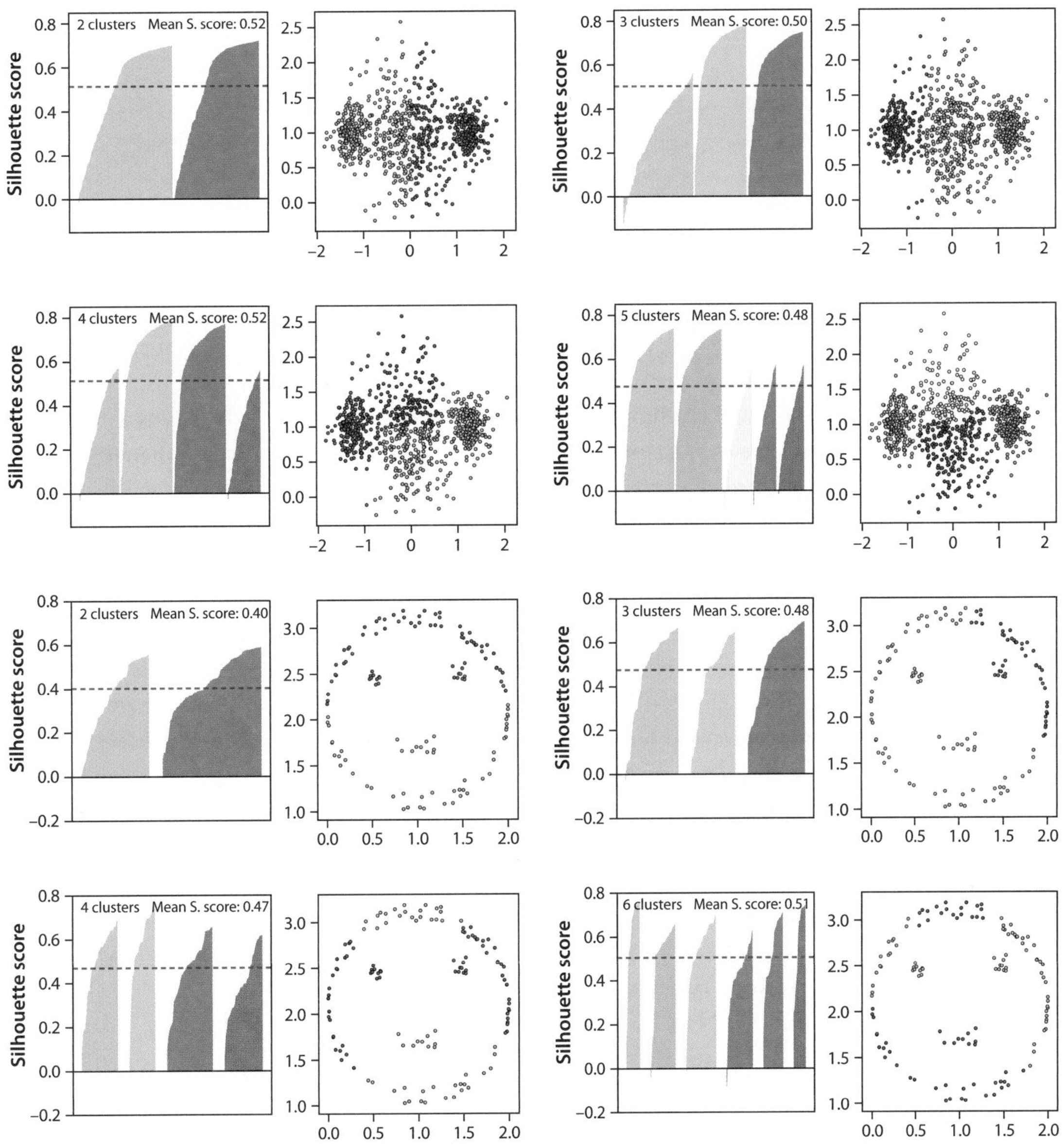


**Figure 7.7: Top four panels: Silhouette scores and cluster assignments from the *k*-means algorithm for 2, 3, 4, and 5 clusters for the "blobs" example. The silhouette score is essentially equivalent for 2, 3, or 4 clusters, but compared to the elbow method, we can use the distribution of silhouette scores to gain a quantitative assessment of overall clustering quality. Bottom four panels: Silhouette scores and cluster assignments for 2, 3, 4, and 6 clusters for the "smiley face" example, again for the *k*-means algorithm.**

```
For each unassigned example x_i:

• Check whether there are at least n_min points within a dis-
  tance of ϵ (that is, whether the sample is a core sample);
• If yes, implement the "expand the cluster" sequence.

"Expand the cluster" sequence:

• Assign all samples within distance ϵ of the current core
  sample to cluster;
• For each newly assigned neighbor x_j that is a core point,
  implement the "expand the cluster" sequence around x_j.
```

Because clusters are built *locally*, on the basis of vicinity among points, clusters of arbitrary shapes can be detected by DBSCAN, and the number of clusters is determined automatically during the process. These might appear as great advantages over $k$-means, but in reality, the number of clusters that are formed (and therefore, the quality of the clustering scheme) is greatly dependent on the proximity threshold $\epsilon$, and in minor measure, on the minimum number of points required for a cluster, as shown in Figure 7.8. The parameter $\epsilon$ is effectively a hyperparameter of the algorithm, and just like the $k$ in $k$-means, it can't be simply determined via cross validation like we do in supervised learning problems. There are some empirical ways to hunt for the best $\epsilon$, which attempt to measure the "typical density scale" of data by plotting its *kNN* distance distribution to estimate $\epsilon$ and are not unlike the elbow method. The OPTICS [Ankerst et al., 1999] algorithm generalizes DBSCAN by incorporating the estimate for the best $\epsilon$, but it is in itself a parametric algorithm and does not work well with the smiley face distribution. The bottom line is that unless we have some knowledge of our data (e.g., some expectations about the shape of the distribution or the number of clusters), clustering algorithms can only provide limited insights.

## 7.3 MIXTURE MODELS

A variation of the global objective function approach is to fit the data to a parameterized model. The parameters for the model are determined from the data, and they determine the clustering. As the name suggests, mixture models assume that the data is a "mixture" of a number of statistical distributions, and they fit the parameters of those distributions. If we use $k$-means as a baseline for comparison, we see some similar limitations (most notably, the number of distributions needs to be specified in advance) but also some significant improvements. Mixture models can accommodate clusters of nonglobular shapes (because they learn about the shape of the distribution, in addition to its center), and even more importantly, *they provide a probabilistic interpretation* of cluster membership, so they are able to retain some information about the degree of confidence in each point's assignment. Thanks to this property, they can be used as a *generative model* to produce new samples that are

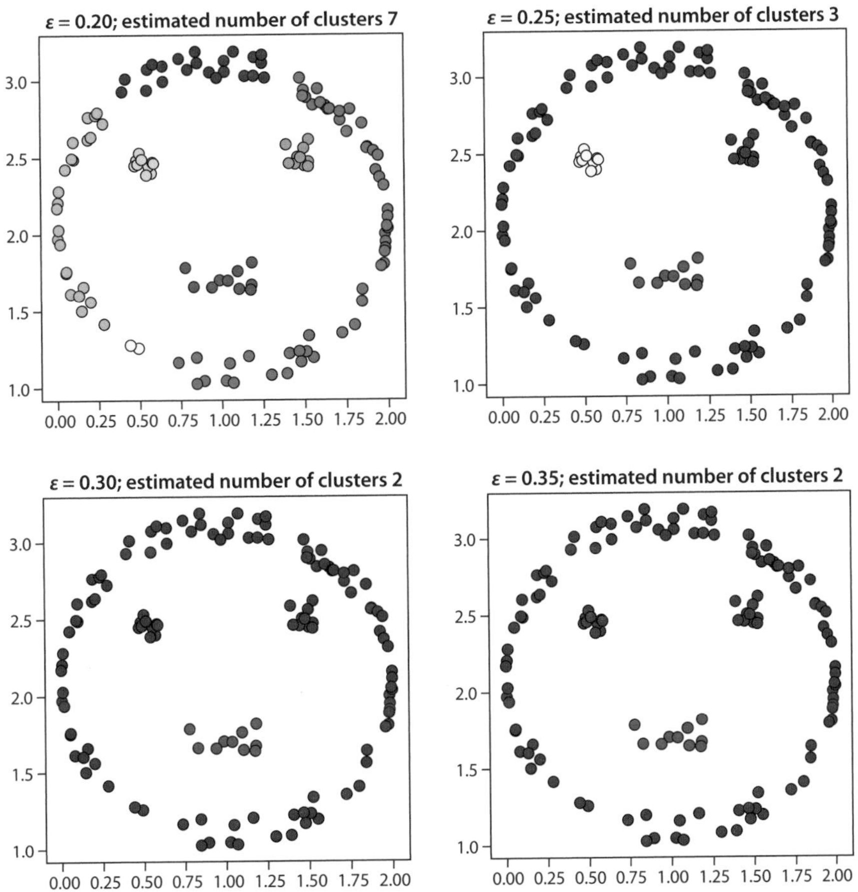


**Figure 7.8: Cluster assignments from the DBSCAN algorithm for a few values of the $\epsilon$ parameter, in increments of 0.05. When the threshold is small, more clusters are created. DBSCAN is very sensitive to the choice of threshold.**

statistically similar to the data. For these reasons, they deserve a special mention in our list of clustering methods, and we'll take a look in detail at one particular case: the Gaussian mixture models.

### 7.3.1 Gaussian mixture models

A Gaussian mixture model (GMM) describes a distribution of data points as a superposition of Gaussian distributions, each of which is determined by a mean value $\boldsymbol{\mu}$ and an $m \times m$ covariance matrix $\boldsymbol{\Sigma}$, where $m$ is, as usual, the dimensionality of the feature space. Just as for $k$-means, we need to specify the number of distributions beforehand; here we will also denote this as $k$. The goal of the GMM clustering algorithm is to use the data to find the parameters $\boldsymbol{\mu}$ and the covariance matrix $\boldsymbol{\Sigma}$ for each one of the $k$ Gaussian mixtures, together with some weights $\tau_i$ that are related to the size of each component and are normalized such that $\sum_{i=1}^{k} \tau_i = 1$. If we can find these parameters, we can calculate the probability that each data point $\mathbf{x}_i$ belongs to component $j$ as:

$$\mathfrak{R}_j(\mathrm{x}_i|\boldsymbol{\tau}, \boldsymbol{\mu}, \boldsymbol{\Sigma}) = \frac{\tau_j N_j(\mathrm{x}_i|\boldsymbol{\mu}^j, \Sigma^j)}{\sum_{i=1}^{k} \tau_i N_i(\mathrm{x}_i|\boldsymbol{\mu}^i, \Sigma^i)}, \tag{7.5}$$

where $N_i(\boldsymbol{\mu}, \boldsymbol{\Sigma})$ denotes the normal distribution with mean $\boldsymbol{\mu}$ and covariance $\boldsymbol{\Sigma}$. This quantity is also known as the *responsibility* $\mathfrak{R}$.

The model assigns a label to each sample, that is, it identifies the Gaussian component to which the data point $\mathrm{x}_i$ is most likely to belong, by choosing the component with the highest responsibility $\mathfrak{R}_j$. Thus the GMM acts as both a description of the overall density distribution of the data and as an unsupervised clustering method, by placing each sample into one of $k$ groups.

In `sklearn`, this process is implemented using the `predict_proba` method, which returns a matrix of size [n_samples, $k$] containing the estimated probability that any point belongs to any given cluster.

From a pseudocode perspective, the GMM works somewhat similarly to $k$-means, as it is also composed of two *expectation-maximization* steps:

1. `Initialize the` $k$ `Gaussian mixture components with some guess at their mean (i.e., the location of the centers of the clusters) and covariance (shape).`
2. `Predict the probability that each point is assigned to a cluster (responsibility), and assign points to the cluster with the highest probability of membership.`
3. `Use the new assignments to compute the new means and covariances for each component, together with the weights` $\tau_i$`.`
4. `Repeat steps 2-3 until some convergence criterion has been reached (e.g., responsibilities do not change significantly).`

GMM works by estimating the parameters of each Gaussian component. It is easy to understand how the corresponding cluster shapes are more flexible than those afforded by $k$-means; if the covariance matrices have no constraints, each cluster is allowed to have an elliptical shape, and the axes of the ellipse can have any orientation. This is qualitatively shown in Figure 7.9. However, the number of parameters to be determined can easily become quite large. There are $k-1$ weight parameters of the vector $\boldsymbol{\tau}$, $k \times m$ means $\boldsymbol{\mu}$, and the $m(m+1)/2$ components for each of the $k$ positive-definite, symmetric covariance matrices $\boldsymbol{\Sigma}$, for a total of

$$n_p = \frac{k}{2}(m^2 + 3m + 2) - 1 \tag{7.6}$$

free parameters in the model. For this reason, one possible option is to restrict the shape of the covariance matrix. The "diagonal" option forces the axes of the covariance matrices to be parallel to the original feature axes while still allowing elongated

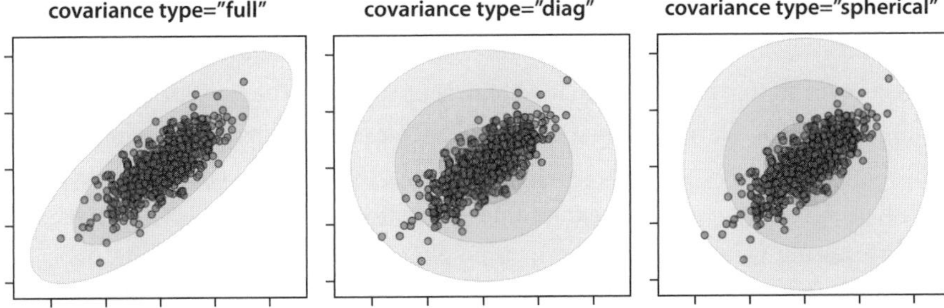


Figure 7.9: Clusters from a Gaussian mixture model will have different shapes when different options for the covariance matrix are selected. The "full" option allows elliptical shapes with arbitrary axes rotation; the "diagonal" option forces the axes of the covariance matrices to be parallel to the original feature axes; the "spherical" option forces clusters to have a spherical shape. Figure inspired by [VanderPlas, 2016].

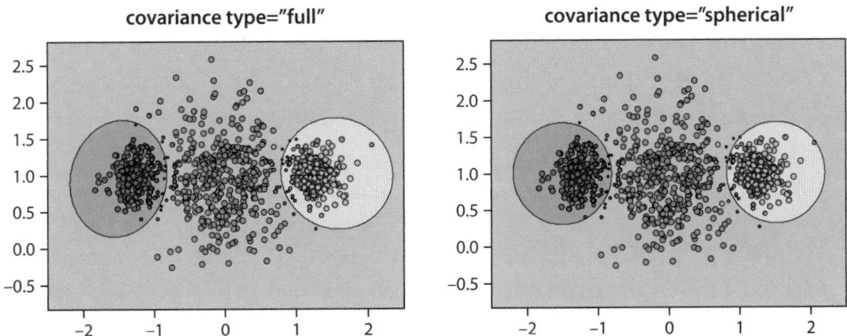


Figure 7.10: Clustering structure from GMMs for the "blobs" distribution introduced earlier in this chapter. The points' colors show their true label (usually unavailable); the background colors correspond to the membership assigned by GMM; the size of each data point is proportional to the confidence in each point's cluster assignment. Smaller points correspond to more uncertain assignments. (Left) Full covariance matrix; (Right) Spherical covariance matrix. The scale is the same in the two panels.

clusters; the "spherical" option forces clusters to have a spherical shape, similar to $k$-means (note that cluster assignments can still be quite different). In all cases, the probabilistic nature of cluster assignments is preserved.

To clarify the differences between GMM and $k$-means, the cluster assignments and the membership probabilities predicted by GMM are shown in Figure 7.10 for a simulated distribution with three clusters. In each panel, the points' colors show their true cluster assignment (usually unavailable); the background colors correspond to the membership assigned by GMM; and the size of each data point is proportional to the responsibility (i.e., the confidence in each point's cluster assignment). Smaller points correspond to more uncertain assignments. As mentioned when discussing Figure 7.5, there is no "perfect" solution for overlapping clusters; however, in GMMs, points that are close to the decision boundaries have less confident attributions (lower responsibilities), which is helpful in understanding the uncertainty in the clustering structure we obtain. The left panel of Figure 7.10 shows the result for the full covariance matrix option, and the right panel shows the result for the spherical covariance matrix. In this case, the distribution of points was almost spherical to begin with, so the shape and orientation of the formed clusters are not too different (note, however, some elongation and rotation in the left panel). It is interesting to note how even in the spherical covariance case, the decision boundaries are very different from those proposed by $k$-means (see right panel of Figure 7.5). The $k$-means algorithm will try to divide the space in areas of similar size to minimize the equally weighted sum of squared distances from centroids, while GMMs are better suited to representing clusters of different sizes and densities.

### 7.3.2 GMMs as generative models

An important characteristic of mixture models is that they have a probabilistic interpretation. This means not only that we are afforded a probability when we use them as a clustering algorithm, but also that we can calculate the *likelihood* of the data, given the model, where the model is a set of distributions with known means and covariances.

This trait is important for two reasons:

1. We can use statistical tools such as the *Bayes Information Criterion* (BIC) [Schwarz, 1978] or the *Akaike Information Criterion* (AIC) [Akaike, 1998] to estimate the number of components present in the data; and
2. We can use the mixtures as a *generative model* to produce new samples that have similar statistical properties to the existing data.

Note that for point 1, as mentioned for the elbow method earlier in this chapter, *the answer that we get is only meaningful in the context of the model that we use.* Using a criterion like the BIC in the context of a Gaussian mixture model will just answer the question "How many elliptical-shaped, arbitrarily rotated components do I need to represent this data?" and not "How many components, or clusters, are in these data?" This is one of the reasons that clustering, and unsupervised learning in general, is often ambiguous: The answer will always depend on the shape of clusters in the data, and using a certain method without prior knowledge of what that will be means that we risk obtaining meaningless results.

Nonetheless, the ability to use mixture models, including GMMs, as a density estimator and thus as a generative model is very useful, as shown by a short demonstration here (you can also find the details in notebook "IntroClustering.ipynb").

Let us use our "smiley face" data. Visually, those data have four clusters, and not surprisingly, trying to fit them with four Gaussian components gives meaningless results, just as we observed with $k$-means in Figure 7.4, although the cluster assignments are significantly different. We can instead use the BIC to select an optimal number of components that can represent our data. The expression for the

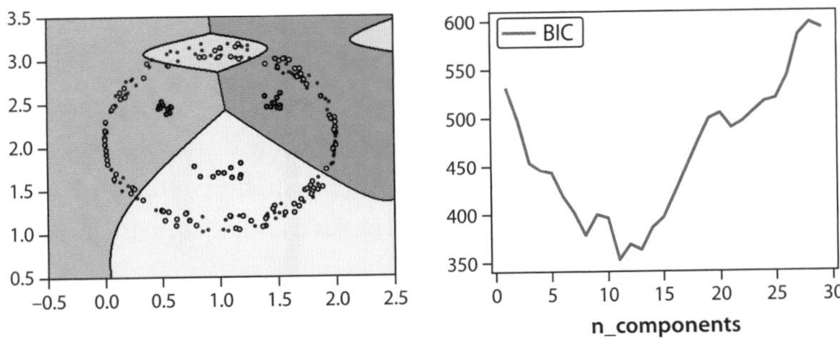


**Figure 7.11: (Left) Despite using the correct number of clusters (4), the GMM is not able to obtain a meaningful clustering on these data, because it is limited to using ellipsoidal cluster shapes. (Right) Bayes Information Criterion, plotted for models with a varying number of components between 1 and 30 and fit using the data on the left.**

likelihood of data, given a GMM model defined by its means, covariances, and weights is:

$$\ln \mathcal{L}(\mathrm{x}|\boldsymbol{\tau}, \{\boldsymbol{\mu}\}, \{\boldsymbol{\Sigma}\}) = \sum_{j=1}^{N} \sum_{i=1}^{k} \left[ \log \tau_i - \frac{1}{2} \log |\Sigma^i| - \frac{m \log(2\pi)}{2} \right.$$
$$\left. - \frac{1}{2} \left(\mathrm{x}_j - \boldsymbol{\mu}^i\right)^T \cdot (\boldsymbol{\Sigma}^i)^{-1} \cdot \left(\mathrm{x}_j - \boldsymbol{\mu}^i\right) \right], \tag{7.7}$$

By "fitting" the GMM model, we effectively derive the configuration (means, covariances, and weights) that maximizes the likelihood. If we repeat this process for different values of $k$, we can determine the number of components that minimizes the BIC:

$$BIC = -2 \ln \hat{\mathcal{L}}_{||} + n_p \ln(N), \tag{7.8}$$

where $\hat{\mathcal{L}}_{||}$ is the maximum value of the likelihood function for the model with $k$ components, $n_p$ is the total number of free parameters in the model, and $N$, as usual, is the total number of samples in our data set. The BIC is composed by two competing terms: The first rewards models with high likelihood, as they provide a good fit to the data, and the second penalizes complex models, with a higher number of free parameters (as of course, the number of free parameters is directly proportional to the number of Gaussian components, as seen in Eq. 7.6). The BIC is already available as an attribute of the Gaussian mixture models in `sklearn`, so we can simply plot it, as shown in the right panel of Figure 7.11. The minimum in the BIC curve tells us that about 10 components can represent our data well; the left panel of Figure 7.12 shows the projected covariances of a 10-component model fit to our data. The gray ellipses (representing the 68, 95, and 99% of probability density associated with each component) provide a good "coverage" of our data points, with the possible exception of the component that accounts for the top right section of the circular distribution and the right "eye" of the face, which are assumed to be drawn from the

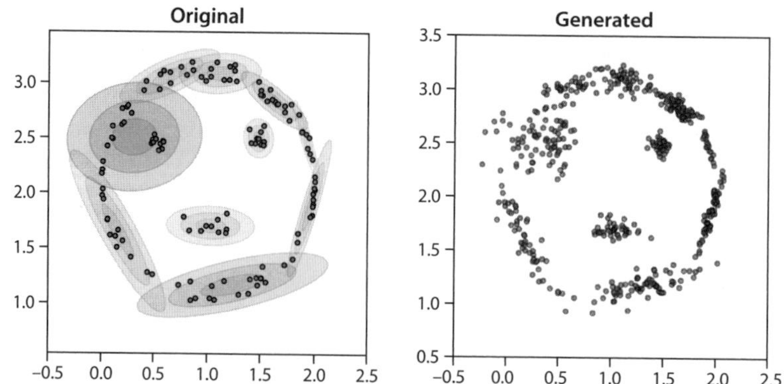


**Figure 7.12: (Left) Projected covariances of a 10-component model fit to the data. The gray ellipses (representing the 68, 95, and 99% of probability density associated with each component) generally provide a good "coverage" of our data points. (Right) Data generated by drawing samples from those 10 components; they are visually similar to the original data.**

same distribution. Looking back at our data, this is understandable, as that region appears to be particularly noisy. Finally, we can go through the exercise of *generating* new data points by drawing samples from those ten components. This is really easy in `sklearn`, using the `sample` property offered by the Gaussian mixture models:

```
Xnew = gmm10.sample(n_samples=500)
```

A scatter plot of the new data, showing the visual similarity between "Original" and "Generated" samples, is shown in the right panel of Figure 7.12.

## 7.4 DIMENSIONALITY REDUCTION

A very common application of unsupervised learning is to reduce the dimensionality of data space, with minimal information loss. It often happens that complex data, like images or spectra, have a redundant structure, and it is possible to explore such redundancy to figure out a different representation that is more efficient. This can be useful for making visualization, data exploration, or model building more agile. In fact, dimensionality reduction is often used as a preprocessing step, before building a supervised learning model. The flip side of this process is that while original features typically have an interpretable physical meaning, the new features created in the efficient representation don't. However, we can retain the mapping between original and new representation, which allows us to at least transform back and forth between the two spaces.

### 7.4.1 Linear PCA

One of the simplest forms of dimensionality reduction is the Principal Component Analysis (PCA). In linear PCA, we seek a small number $k$ of vectors in the original feature space so that every element of our data set can be expressed as a linear combination of those $k$ elements, with the minimum possible information loss. The new representation will be obtained by projecting of the original samples onto the smaller-dimensional space described by these elements, which are the *principal components.* Just as for clustering, the loss function is tightly connected to a notion of similarity: A typical loss function is the average squared Euclidean distance between the "original" and "new" representation of the same object (which is, by the way, our old friend, the MSE). Note that the elements that constitute the "basis" in the new representation don't need to be elements of the original data set (just like in clustering, the centroids can be interpreted as representative elements of their cluster, but they are not bound to be part of the original sample).

It can be shown that, for a fixed dimensionality of the projection space, the principal components can be obtained iteratively as the set of orthogonal components that *maximize the projected variance of the data;* maximizing the projected variance is equivalent to minimizing the MSE of the projected data (see, e.g., these notes[2] for a proof). What does this mean in practice? It means that to find the principal components (PCs), we can do the following:

- Identify the first PC as the direction of greatest variability in data; in other words, data points are most spread out when projected on the first PC, compared to any other direction. The second PC is the next direction of greatest variability, *orthogonal to first PC*. The third PC is the next direction of greatest variability, *orthogonal to first and second PC*.
- Continue until satisfied with the amount of variance in the data that has been captured.

The additional information (variance) of the data decreases as we add PCs, until we reach the number of independent dimensions in the data, so we can decide on our own the trade-off between obtaining a stronger dimensionality reduction with fewer components, or capturing a higher fraction of the variance.

The mathematical formulation of the pseudocode above is quite easy. *As long as the data are centered* (i.e., the average of the data matrix $X$ is 0 for every column), it follows that

1. The principal components are the eigenvectors of the covariance matrix $X^T X$ of the data.
2. The ordering is given by the magnitude of the eigenvalues, with the first PC corresponding to the largest eigenvalue, and so on.

The number of eigenvectors $\alpha_i$ (each one of length $m$) that we keep determines the dimensionality of the projection space. If we stack them together *column-wise,* we find the *projection matrix,* which will have as many rows as the original

2 https://www.stat.cmu.edu/~ryantibs/advmethods/shalizi/ch17.pdf

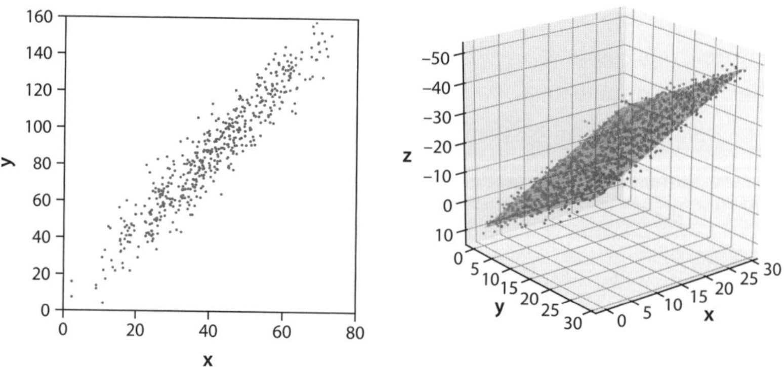


**Figure 7.13: Two simple examples in which the data live in two (left panel) and three (right panel) dimensions, but it's easy to see that one- or two-dimensional projections, respectively, are sufficient to encompass most of the variance.**

data set dimension ($m$), and as many columns as we want to keep ($k$). The new representation is found by multiplying each data point by the projection matrix:

$$\mathrm{x}_i(\mathrm{new}) = \mathrm{x}_i^T \cdot P; \quad P = ((\alpha_1), \ldots.(\alpha_k)). \tag{7.9}$$

The result is a vector of length $k$, which encodes the projection of the data point $\mathrm{x}_i$ onto the subspace of the $k$ principal components.

Figure 7.13 shows two simple examples in which the data exist respectively in two and three dimensions, but it's easy to see that one- or two-dimensional projections are sufficient to encompass most of the variance. Intuitively, in the case of the two-dimensional data, using a "rotated" axis as a coordinate instead of the original $x$ and $y$ coordinates would be beneficial, because the variance of the data would be almost completely captured by a single coordinate in the system of the rotated axes, so that we could identify a data point with one number instead of two.

In more realistic cases where the dimensionality of the original space is large, what is the dimensionality reduction obtained through the PCA algorithm?

If we use $k$ principal components for a data set with $N$ instances and $m$ features, the new representation of the data set will be encoded by the $k$ principal components ($k$ objects with $m$ features), which is to say the projection matrix from Eq. 7.9, plus $k$ coordinates for each of the $N$ objects.

For example, imagine that the original data set is made up of 10,000 $64 \times 64$ images. In this case, $m = 4{,}096$, $N = 10{,}000$. The original data set size is $m \times N \sim 4 \times 10^7$. If we can describe each image well enough as a superposition of, say, 20 eigen-images, then the size of the new data set is $20 \times 4{,}096$ for the projection matrix containing the feature values of each eigen-image, $+ 20 \times 10{,}000$ (the 20 coefficients for each of the original images) $\sim 3 \times 10^5$. This trick reduced the dimensionality by a factor of over 100. More generally,

$$\text{Size (new data)} = k \times m + k \times N = k \times (m + N) << m \times N, \quad \text{if} \quad k << m, N.$$

Not surprisingly, we obtain a significant dimensionality reduction when the number of principal components retained, $k$, is much smaller than the number of original features and the number of objects in our data set.

The final question is: When is PCA expected to work? The answer is that for this process to be effective (cause minimal information loss), *there needs to exist a lower-dimensional representation of the data that can be represented by a linear transformation of the original space.* If the variance of the data set is uniformly distributed across all features, then the loss of information in any projection will be significant. Luckily, in most large-sized inputs used in physics or astronomy, such as maps, spectra, or images, there tends to be a lot of redundancy; for example, we can think of images that have a source at the center and then a certain portion of background around it. Background pixels are very similar for all objects in the data set, so that the variability of the data set associated with those features is small, and we can project images over a smaller-dimensional space without sizable information loss.

### 7.4.2 PCA for spectra

Let us look at a worked example inspired by this exercise.[3]

In this exercise, we create a PCA representation of 4,000 spectra from the Sloan Digital Sky Survey. Each of the spectra has 1,000 measurements, corresponding to the brightness of the spectrum at 1,000 different wavelengths. The dimensionality of this initial data set is thus $4{,}000 \times 1{,}000 = 4 \times 10^6$.

We want to explore the possibility of capturing each spectrum with considerably fewer components (e.g., we could try to create a PCA representation with 50 components). It is worth noting that in standard PCA, *if the true dimensionality of the data is larger than the number of components,* the representation is unique, because the PCs are ordered according to decreasing eigenvalues of the data's covariance matrix. This is not strictly true if the covariance matrix also has zero-valued eigenvalues. Thus we can create a bigger representation, use it to test the information loss for various dimensionality reductions, and then select the number of components we want.

In this case, as mentioned earlier, we first need to make sure that the data are centered; note that in this unsupervised learning process, we don't need to distinguish between train and test sets:

```
scaler = preprocessing.StandardScaler(with_std = False)
Xn = scaler.fit_transform(X)
```

3 https://ogrisel.github.io/scikit-learn.org/sklearn-tutorial/tutorial/astronomy/dimensionality_reduction.html#sdss-spectral-data

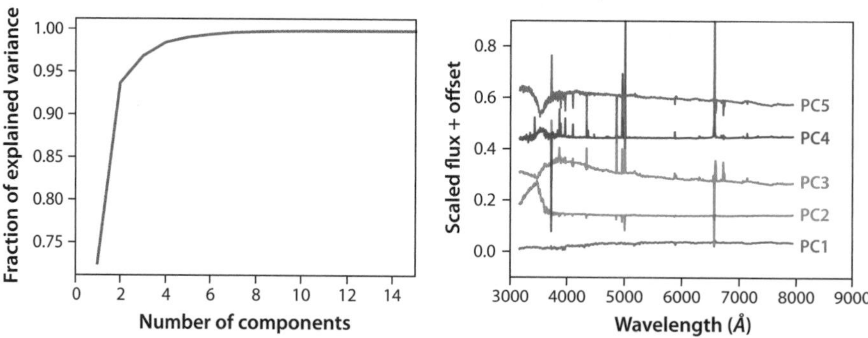


**Figure 7.14: (Left) The cumulative variance explained as a function of number of principal components retained in the representation, for the first 15 PCs. (Right) Plotting the first few principal components can help develop some intuition of "independent modes" in the data. Note that the difference between this plot and the original inspiration quoted in the main text comes from the different standardization technique applied to the data.**

We can then create our PCA decomposition with 50 components on the centered data; this calculates the eigenvectors of the covariance matrix, sorts them according to decreasing eigenvalues, and creates the projection matrix, which is $1{,}000 \times 50$ in this case. Finally, we create the new data set *X_proj_50*, which has 4,000 objects and 50 features per object:

```
pca_50 = decomposition.PCA(n_components = 50, random_state = 0)
X_proj_50 = pca_50.fit_transform(Xn)
```

Once we have our new representation, we can inquire about its efficiency. As mentioned earlier, the whole process of finding PCs and projecting onto them relies on the idea of *maximizing the amount of variance captured*. For this reason, it might be helpful to look at the amount of "explained variance," a property of the PCA estimator (`pca_50.explained_variance_ratio_`) that shows how much of the variance in the initial data set is explained as a function of the number of principal components that are retained, and plot it, as shown in the left panel of Figure 7.14. It is evident that the vast majority of the variance is encapsulated by the first few components (perhaps four or five; those are shown for reference in the right panel of the same figure). The variance added by further components is very small; the cumulative variance explained by the first five components is 99.1% of the total, and after the first ten components it is 99.6%. These numbers seem to imply that we can use just five or ten components to describe this set of spectra, without significant information loss. However, as we will see in the next section, making these assumptions may be quite dangerous.

### 7.4.3 Evaluation metrics for unsupervised learning

Because by nature we are suspicious scientists, we can try to explore a bit further the robustness of the PCA projection with just a handful of components. Another

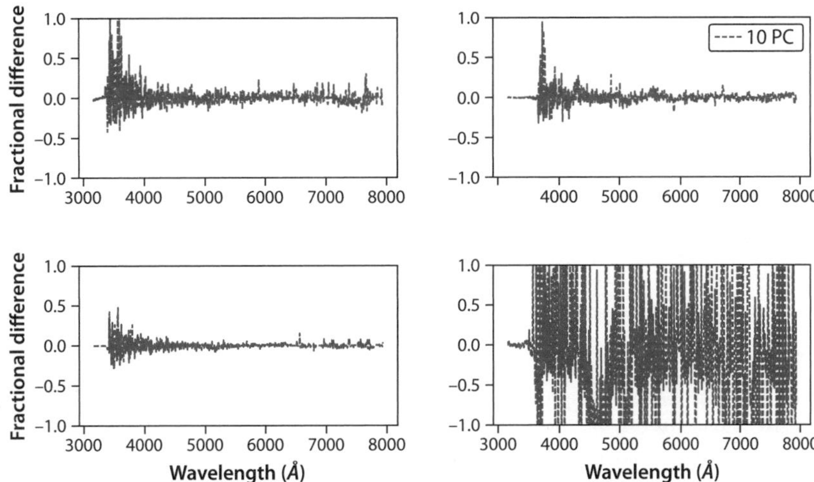


**Figure 7.15: Fractional difference ($(X_{new} - X_{orig})/X_{orig}$) between original and new representation, for the PCA projection with 10 components. The y axis corresponds to a (−100%, 100%) difference. There are significant differences, which are particularly marked for some spectra and in proximity to "spiky" features.**

way to assess the efficacy of the reduced-size representation is to look for additional measures of similarity between the original samples and the new, "projected" ones. As an example, let us retain 10 components, and "project back" the spectra (by inverting the projection process) to their original space:

```
pca_10 = decomposition.PCA(n_components = 10, random_state = 0)
X_proj_10 = pca_10.fit_transform(Xn)
Xrec_10 = pca_10.inverse_transform(X_proj_10)
```

This allows us to plot the fractional difference between the original and reconstructed spectra ($(X_{new} - X_{orig})/X_{orig}$) for the first few objects, shown in Figure 7.15. This plot reveals some interesting facts:

- Some spectra are better represented than others (e.g., compare the fractional differences in the first three vs the fourth panel), as the explained variance refers to the entire data set. This is unsurprising but worth noticing.
- The fractional difference in some regions of the spectra tends to be larger (e.g., in our case, the spectral reconstruction seems to be less reliable at blue wavelengths, $\lambda < 4000$ Å).
- "Spiky" features, such as emission or absorption lines, are not well portrayed in our low-dimensional representation. This is again unsurprising, because the difference in very narrow regions of feature space (e.g., a single wavelength) does not affect the variance too much, but again it is worth noticing.

Do we feel equally confident about the quality of the PCA representation with 10 components after seeing the plots in Figure 7.15? The answer, of course, *depends on the use we want to make of these reduced-size data.* Do we want to just show our collaborators what typical spectra look like, and what their typical shapes are? Probably yes. Do we want to use the data to derive some physical properties of galaxies that rely on indicators such as the strength of emission and absorption lines? Probably not. The problem with using unsupervised methods like PCAs for dimensionality reduction is that *methods that are not specialized for our goals are unlikely to be ideal.* This is a problem that I have encountered often, and in a way, it resembles the challenges we saw in supervised learning, where the evaluation metrics chosen to *build* a model and to *evaluate* a model may be different and deserves a lot of thought.

What can we do to alleviate this problem? A good practice, of course, is to carefully select the most suitable evaluation metric to assess the quality of the DR method. For example, in the previous example, we could decide that we would not accept a fractional difference between "original" and "new" spectra beyond 10% anywhere in the spectra and then figure out how many components are needed to satisfy this new quality assessment measure.

We can also look at more powerful (i.e., nonlinear) dimensionality reduction techniques, which would give us a chance to obtain a better representation with fewer components, and consider *supervised* dimensionality reduction techniques, as described in the next two sections.

### 7.4.4 Kernel PCA

One of the main limitations of linear PCA is that... well... they are linear! Any new representation is made of linear combinations of elements of the original feature space, which of course narrows the range of variability that can be expressed with a few components and induces a certain amount of information loss.

Luckily, there are many other possible representations that "go beyond" linearity, perhaps decreasing the interpretability of mapping but increasing the flexibility of the space. One notable one is *Kernel* PCA [Schölkopf et al., 1999], which makes use of the same "kernel trick" introduced in Chapter 4, where we talked about SVMs and implicit transformations onto higher-dimensional spaces.

Kernel PCA employs the same strategy: Once we choose the functional form for our kernel (e.g., polynomial, or Gaussian), we effectively gain an upgraded orthonormal basis to represent the data *without ever specifying the mapping explicitly.* In fact, it can be shown that the principal components of Kernel PCA are the eigenvectors of the kernel matrix (rather than the eigenvectors of the data covariance matrix); as a result, they are nonlinear combinations of some vectors in the original feature space.

Figure 7.16 illustrates how Kernel PCA works; we use the `KernelPCA` class that is available in `sklearn`. The original data involve two distinct populations, the outer circle and the inner circle, which are colored blue and green to identify

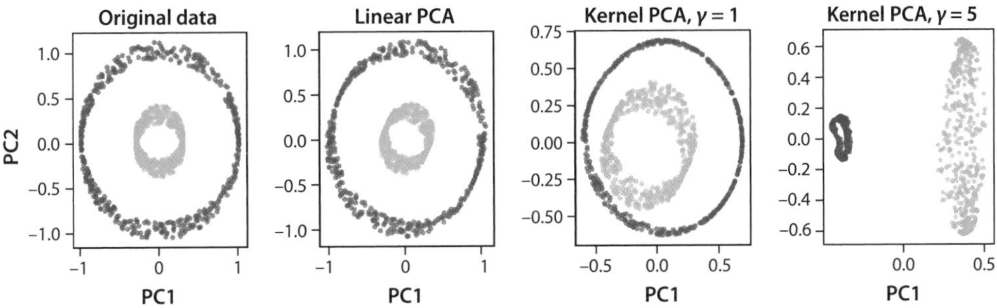


Figure 7.16: **The original data, shown in the leftmost panel, live in two dimensions and allow us to distinguish between the two populations (characterized by different colors for illustration purposes). Linear PCA (second panel from left) can't engineer a transformation to a lower-dimensional space that keeps the information about population membership (i.e., color), but Kernel PCA can, provided that the correct value of $\gamma$ is chosen. For $\gamma = 5$, projecting onto the first principal axis (the *x* axis of the rightmost panel) would allow one to retain the original information. Figure inspired by the documentation in [Pedregosa et al., 2011].**

them more clearly. It's easy to distinguish them in the original data space, shown in the left panel of the figure, as they are spatially well separated. We can ask: Is there a lower-dimensional representation of the data (in this case one-dimensional) that can retain the information about the color, and hence, the ability to distinguish between the two sets? It is clear that there is no linear transformation (i.e., rotation of the axes) that can do so. In fact, if we apply linear PCA to those data, the new representation along the PC1 and PC2 axes is remarkably similar to the original, as shown in the second panel. If we were to project the data onto the first component (imagine "squeezing" all the points vertically to a y coordinate of zero), we would remove a lot of information, as shown by the fact that we wouldn't be able to distinguish between the green and blue sets of points any more, while this task was easy in the original representation. However, if we calculate the first two principal components for a Kernel PCA with a Gaussian kernel, we have much more freedom in the mapping, as shown in the third and fourth panels of Figure 7.16. Recall from the illustrations in Chapter 4 that very small values of $\gamma$ resemble a linear kernel, while higher values become progressively more nonlinear. In our case, we can see that for $\gamma = 1$, the principal components are significantly nonlinear, and for $\gamma = 5$, we are actually able to retain the color information about the two populations even if we project onto the first principal component alone (the x axis of the rightmost panel in the figure). This doesn't, of course, show that the information loss is zero; it just shows how more information is lost when we use linear PCAs vs the Kernel version.

Nonlinear dimensionality reduction techniques are in general very powerful, but there are important caveats in using them. One is that, as shown above, they rely heavily on hyperparameter optimization, which should be carried out through cross validation. How does CV work in unsupervised learning? Usually, the success metric for selecting hyperparameters is some measure of information loss. In our earlier example, we looked at the fractional difference between the original spectra and those reconstructed by inverting the process of mapping the data to the lower-dimensional space. This gave us an idea of how much of the original information was still available in the low-dimensional representation. Other choices could be the MSE (or Euclidean distance, equivalently) between original inputs and "reconstructed" ones; this idea is the basis for *autoencoders*, powerful representation mappings that we'll discuss in the next chapter. The other caveat is that nonlinear representations don't always work, and they lack some of the interpretability of linear methods. For example, you are invited to check for yourself (in the exercise section) that using Kernel PCA is not very helpful in mapping spectra to a lower-dimensional domain.

## 7.5 APPLICATION: HYPERSPECTRAL IMAGES ANALYSIS

One of the most powerful applications of clustering and dimensionality reduction techniques is to visualize and understand data, especially data that are visual in nature, such as images. This field takes the name of *computer vision*.

Let us apply some of the tools introduced in this chapter to analyze data from the remote sensing domain: hyperspectral images. Hyperspectral images are data that contain both spatial (x-y) and spectral (wavelength $\lambda$) information. For each pixel in an image, information about the spectral emission of that pixel (namely, a finely sampled chart of the emission as a function of wavelength) is collected, as shown in Figure 7.17. This is useful, for example, in aerial surveys, to identify different types of terrain based on their spectral signature and location. Note that typical RGB images are also a form of superposition of emission charts in various bands, where of course the bands are very wide (they record the average emission in the red, green, and blue regions of the EM spectrum). However, the single-pixel information from RGB images is usually not compelling enough, so images typically are processed as a whole (every pixel is a feature, and every image is an instance), while in hyperspectral images, every spectral measurement is a feature, and every pixel is an instance.

### 7.5.1 Data set description

Consider a "classic" data set in the remote sensing domain, the Pavia data set collected by the European Reflective Optics System Imaging Spectrometer (ROSIS-3) satellite. The data contain two images, one taken over the center of the Italian city of Pavia, and one taken over the University area. We use the "Pavia Centre" data set, which consists of one composite image of $1{,}096 \times 715$ pixels, with 1.3 m spatial resolution; for each pixel, a record of its spectral emission in 102 bands between 430 and 834 nm is available. The data set was downloaded from the website of the Grupo

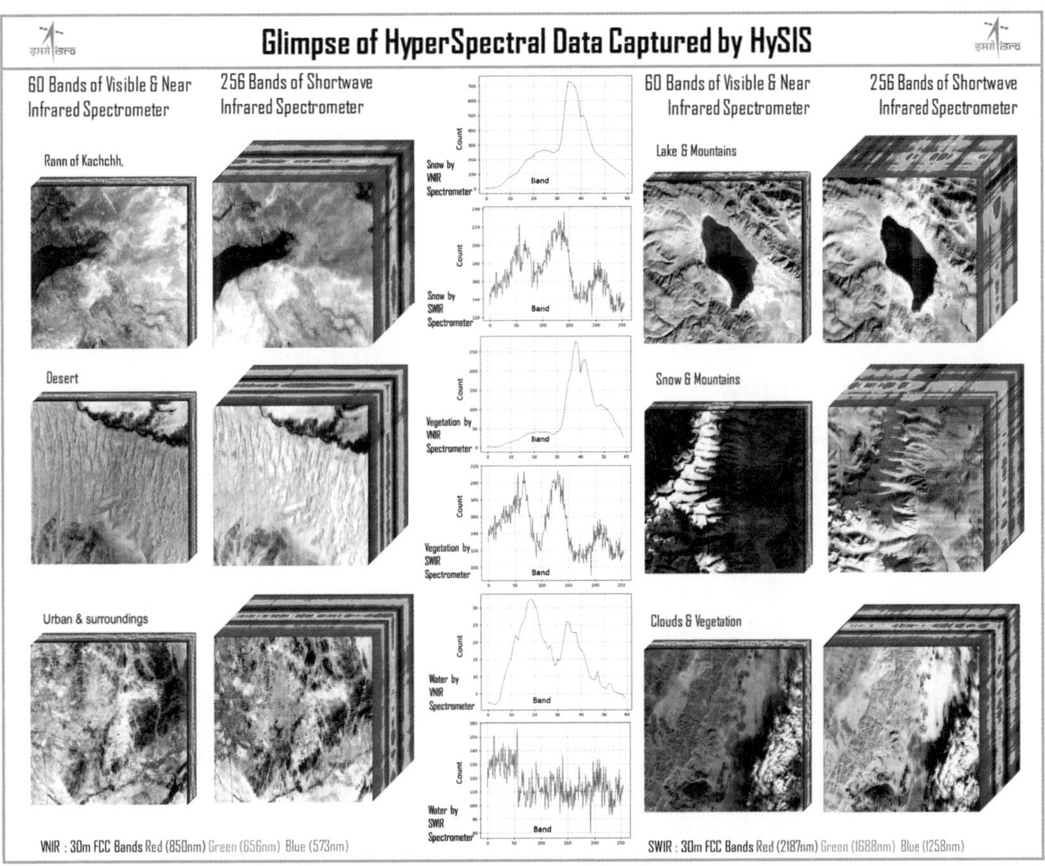


**Figure 7.17: The first results from the Hyperspectral Imaging Satellite show the hyperspectral signatures of different types of terrain. Figure from the Indian Space Research Organization (https://www.isro.gov.in/).**

de Inteligencia Computacional de la Universidad del País Vasco.[4] For our analysis, each pixel is an instance, and each instance contains 102 features. This is clearly not optimal, because by treating each pixel as an independent entity, we neglect to include spatial information (e.g., the fact that adjacent pixels are more likely to belong to the same class). Labeling information is also available for about half of the image pixels; labels correspond to one of nine classes, indicating a land-cover type. The labels and class membership are shown in Table 7.1. Note that because labels are available, we could approach this problem with supervised learning techniques; however, here we are trying to understand how different clustering and data reduction techniques can help explain and analyze data. An RGB composite of the Pavia Centre scene is shown in Figure 7.18.

4 https://www.ehu.eus/ccwintco/index.php/Hyperspectral_Remote_Sensing_Scenes

Table 7.1: **Labels and class membership for the nine types of land cover in the data set.**

| Label | Class | Membership |
|---|---|---|
| 0 | Unlabeled | 635488 |
| 1 | Water | 65971 |
| 2 | Trees | 7598 |
| 3 | Asphalt | 3090 |
| 4 | Self-blocking Bricks | 2685 |
| 5 | Bitumen | 6584 |
| 6 | Tiles | 9248 |
| 7 | Shadows | 7287 |
| 8 | Meadows | 42826 |
| 9 | Bare Soil | 2863 |

### 7.5.2 Can we find clusters for classes?

The question that clustering algorithms can answer is whether it is possible to find in the data clearly separable groups of objects that share some similarity. In this case, we can assume that pixels that correspond to the same land-cover type (e.g., water or trees) will also have a similar spectral signature. By clustering the data in feature space, we can hope to recover the original classes.

Let us start the investigation with the simplest clustering algorithm, *k*-means, and "cheat" a little by indicating that we know a priori how many clusters we are looking for. Pixels with the label "0," which are unlabeled, will be discarded for this exercise, because we will be using the labels afterward to interpret our results. You can follow the narrative of this section in the notebook "PaviaCentre_Exploration.ipynb." After masking the unlabeled pixels and scaling the data, we can generate a predicted class distribution in a couple of lines of code; as usual, we fix the random seed to ensure reproducibility:

```
kmeans = KMeans(n_clusters = 9, random_state=0)
clusters = kmeans.fit_predict(maskdata_scaled)
clusters = clusters + 1
```

The last line serves to match the range of "true" and "predicted" labels. We can look at the distribution of objects in clusters and compare it to the original class distribution to get a general idea of whether the clustering was successful; this comparison is shown in Figure 7.19. Note that unlike those produced by a supervised classification algorithm, the labels produced by clustering are only meaningful in a relative sense; objects in the same group will share the same label, but each group may be assigned any particular label. For example, in both distributions, there is a

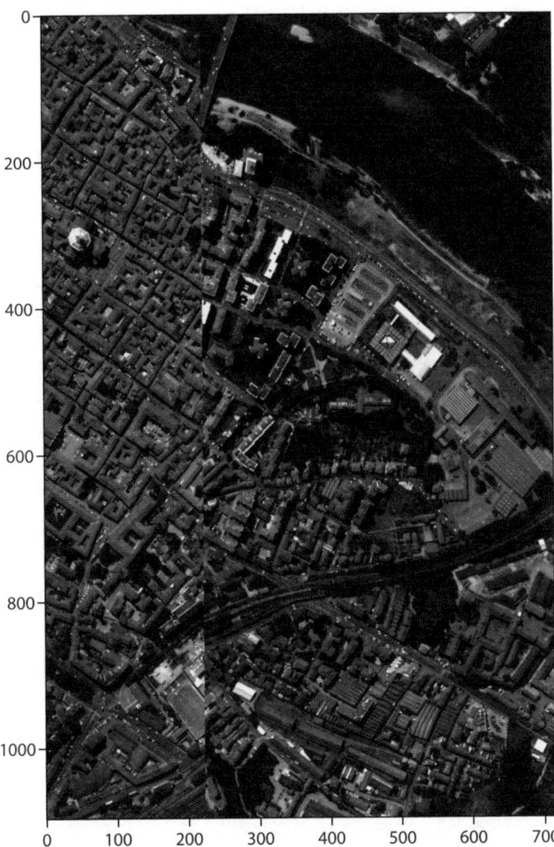


**Figure 7.18: False RGB color composite of the Pavia Centre image. Note that the image is a composite of two different areas that are simply patched together.**

clearly a largest cluster (which has the label 1, corresponding to "Water" in the "true" class distribution, while it happens to have the label 2 in the clustering results), but the second largest cluster, which is present in the ground truth and corresponds to label 8 (Meadows), does not seem to appear in the clustering results, suggesting that objects with that label have been split among different clusters.

Because we can't compare the two distributions directly, to evaluate the quality of our clustering result, we need a metric that measures the similarity of the two groupings and is *invariant to permutations*. There are several candidates: The one we use here is the *Adjusted Rand Index* (ARI).[5] This metric is also useful for estimating the similarity of clustering results obtained by different algorithms. The ARI varies between 0 (which corresponds to the similarity expected for a random grouping) and 1 (indicating perfectly equivalent groupings). In this case, the ARI is $\sim 0.77$. As usual with measures of correlation or similarity, there is no direct interpretation of this number; we can think of it as "not bad."

5 https://en.wikipedia.org/wiki/Rand_index

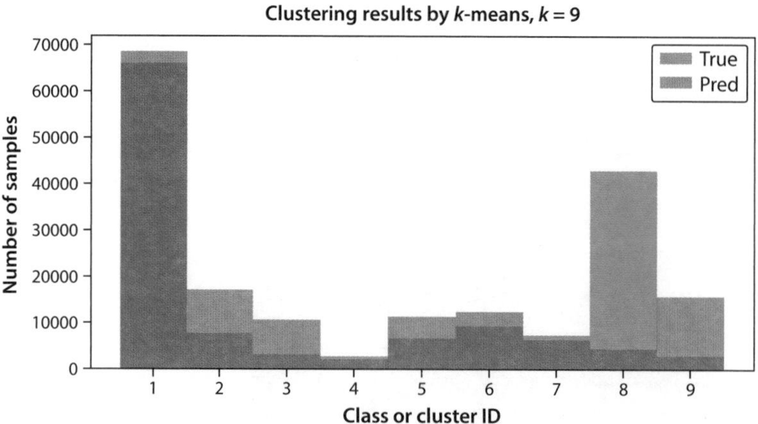


**Figure 7.19: The true distribution, based on the ground truth class labels (magenta), and the distribution in clusters (green) predicted by *k*-means with $k = 9$.**

If we had a way to meaningfully associate each cluster with a label, we could gain further insights into the clustering results by using supervised learning metrics and comparing the ground truth with the predicted labels. We can attempt to do so by calculating the centroids of each class from the ground truth labels, and "mapping" the centroid of each cluster found to the closest ground truth centroid. For example, objects in cluster 2 are mapped onto class 1, confirming our earlier intuition that the pixels belonging to the largest class ("water") have been identified correctly. Note that this procedure doesn't ensure that each cluster is mapped onto a class; in fact, no centroid is mapped to class 3 (asphalt) or class 9 (bare soil). Note that these are two of the three least populated classes; we will return to this argument in the next section.

We can now "redefine" the labels predicted by the clustering process using the mapping we have obtained and use them to compare the ground truth of the image with the predictions of *k*-means. The result is shown in Figure 7.20.

This notation also allows us to calculate metrics such as the accuracy, which is the fraction of (labeled-only) pixels that are identified correctly. In this case, we obtain 80.7%, which is actually competitive with published results for more selective data cuts [Murphy and Maggioni, 2018; Yadav et al., 2019].

### 7.5.3 Dimensionality reduction for visualization purposes

To understand the behavior and decision-making of our clustering scheme, or even just to develop intuitions about the data, it is often useful to use visualization techniques. Dimensionality reduction tools can come to the rescue here: It would be hard to plot data in 102 dimensions, such as these data, and we'd like instead to find a way to capture essential features in two or three dimensions.

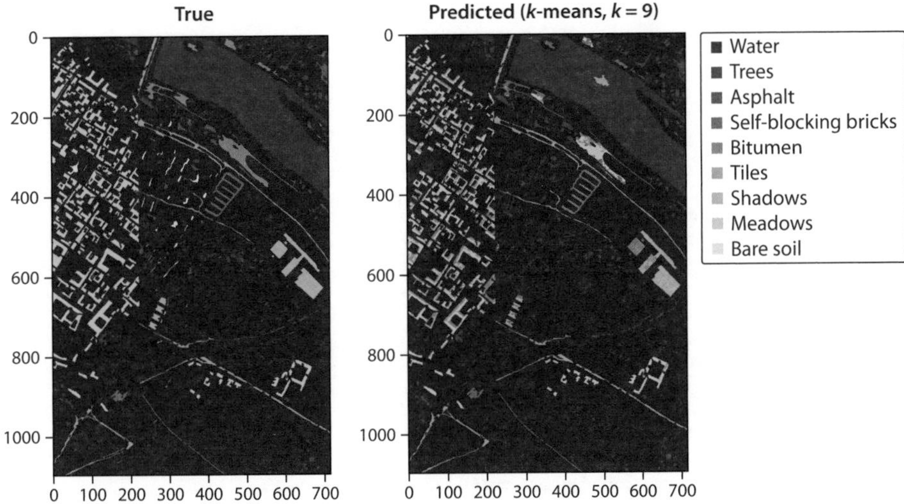


**Figure 7.20: Ground truth labels (left) vs clustering predictions after the centroid mapping procedure described in Sec. 7.5.2 (right). The majority of pixels are predicted correctly (with an accuracy score, excluding unlabeled pixels, of 81%), but the images show some differences; for example, orange-ish points, corresponding to label 9 (bare soil) are never found in the predicted labels image. There are also spurious structures that are only present in the predicted image, as well as salt-and-pepper noise in various locations.**

The first algorithm to reach out to, in my humble (narrator: not so humble) opinion, will always be linear PCAs. If the shapes of data (in this case, the spectral signature of each pixel) are reasonably well behaved, without spiky features, we can expect that most examples in the data set will be well approximated by a linear combination of a few PCA components, and that the Euclidean distance between spectra will be a useful measure of similarity.

In this case, we build a five-dimensional PCA representation of the data. The fraction of cumulative variance expressed by the first two and three components is 95 and 98%, respectively, which is encouraging. We will use the first two components for visualization purposes. We can begin by simply plotting all the points, colored according to their ground truth labels; this is shown in Figure 7.21. From the plot, we can understand why no elements were assigned to class 3 (asphalt): It is almost contiguous to class 2 (trees). It may seem less clear why class 9 (bare soil) is not picked up as a cluster, given its decent separation from the water class, but it's good to keep in mind that the water class contains many more elements, and the cost function of the k-means algorithm will favor incorporating smaller clusters into larger ones. Also remember that we are looking at a two-dimensional projection of a 102-dimensional data set, and so we can't extract *all* information from this plot.

Finally, a useful piece of information from this visualization is that, at least in the two-dimensional PCA space, the distributions of objects in different classes don't

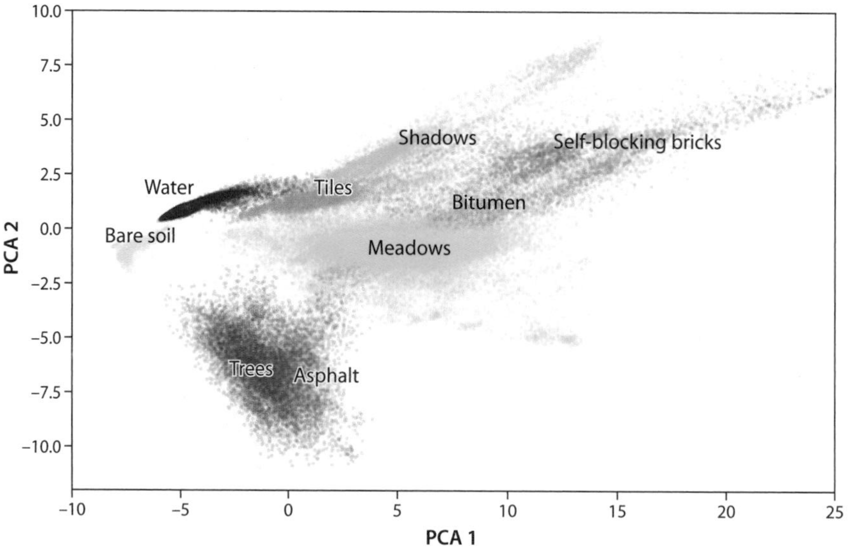


**Figure 7.21: Visualization of all data points, colored according to their ground truth labels, in the space of the first two PCA components.**

appear to be spherical. This reveals a possible limitation of the $k$-means algorithm for clustering of these data.

Another useful exercise consists of plotting, in the same two-dimensional PCA space, the centroids of the ground truth classes, as well as the centroids identified by $k$-means, as shown in Figure 7.22. This confirms some of the intuitions described above. Classes with large membership (e.g., water, meadows) are recognized clearly, and centroids created by the clustering algorithm largely coincide with the "true" class centroid; others, such as tiles, are also identified quite well. For others, confusion prevails; only one cluster (i.e., one centroid) is created for the asphalt and trees classes, and for the water and bare soil classes; there are a few incorrectly inferred centroids in the central region of the plot, where the largest amount of overlap among classes occurs.

### 7.5.4 Estimating the number of classes

We have used this example, where the ground truth labels are actually available, to understand better how clustering and dimensionality reduction algorithms can be used to visualize and understand patterns in data. In the previous sections, we used the information about the true number of classes in the data (9) as an input to the $k$-means algorithm, but typically, this is not available. A more meaningful question to ask is: *How many clusters emerge from the data alone?* This might look like a pointless exercise, but I actually find it very important, because we can answer the question: *Would data like these (these features, this data set size, etc.) allow us to distinguish between these types of objects?* Additionally, even when the number of original

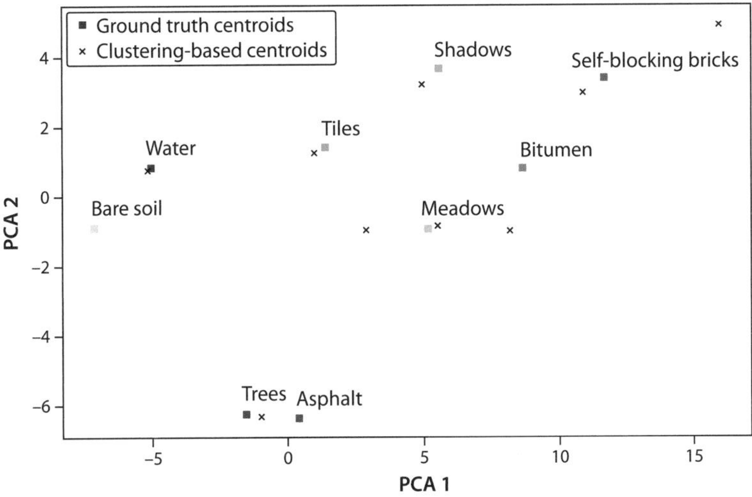


**Figure 7.22: Visualization of ground-truth based class centroids, and centroids inferred by the *k*-means algorithm, in the space of the first two PCA components.**

classes is known, as in the previous case, we have no guarantee that the clusters we find will match the original distribution. For example, the fact that the Trees and Asphalt classes are merged by $k$-means into one cluster is not necessarily an indication of a failure of the $k$-means algorithm; it also points out that the spectral signatures of these classes are similar, at least if we use Euclidean distance as a measure of similarity. Therefore, we might need to collect different features, or improve our definition of similarity measure, to discriminate between them.

Among the strategies we can employ to figure out how many clusters emerge from the data are the elbow method, the silhouette score, or the BIC if we use a generative model like Gaussian mixtures. Note that the answer will depend on the clustering algorithm as well: If we employ $k$-means, we are effectively asking "how many convex, spherical, similarly sized clusters are in the data?", while if we use GMMs, we are asking "how many convex, possibly rotated and elongated, clusters are in the data?"

In the lecture notebook "PaviaCentre_Exploration.ipynb," we use the elbow method for $k$-means and (modulo the usual ambiguity in finding the elbow in the plot) conclude that $k = 5$ could be the chosen answer. While this may seem disappointing at first, because our data set has nine classes, the Adjusted Rand Index of the distribution predicted by this algorithm (and compared to the ground truth) is $\sim 0.74$, not very distant from the result we had ($\sim 0.77$) for $k = 9$. The similarity of distributions is a better indicator of whether the clustering was successful, compared to the number of clusters found; what we are observing here is that the clusters that emerge from the data are not *terribly* dependent on previous knowledge of the number of classes. Overall, I'd say that this is good news.

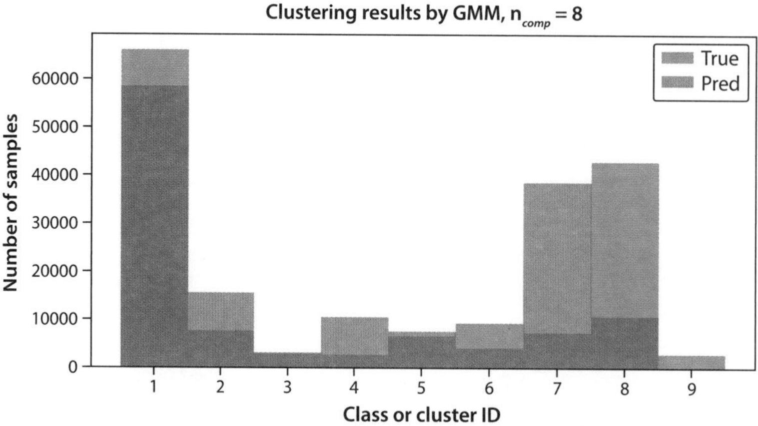


**Figure 7.23: The true distribution, based on the ground truth class labels (magenta), and the distribution in clusters (green) predicted by a Gaussian mixture model with 8 components; the number of components was tuned using the BIC. This distribution is a better match to the original than the one found by $k$-means and shown in Figure 7.19, thanks to the higher flexibility of the GMM model and the availability of a large number of training examples.**

As a more instructive exercise, let us see what happens when we use the Gaussian mixture model to create clusterings. We can feel optimistic that we might obtain a better clustering result, because GMMs are more flexible than $k$-means in terms of allowing clusters of different shapes and orientation. Their main weakness is that they require fitting a large number of parameters (especially if we want to use the full covariances), but because we have a large data set with $\sim$ 150,000 examples, we can be optimistic about our ability to fit those parameters.

We create several clusterings, for a varying number of components between 3 and 10:

```
n_components = np.arange(3, 11)
models = [mixture.GaussianMixture(n, covariance_type='full',
random_state=0).fit(maskdata_scaled) for n in n_components]
```

The preferred number of components is the one that minimizes the BIC:

```
print(np.arange(3,11)[np.argmin([m.bic(maskdata_scaled) for m in models])])
```

In this case, the GMM "picks" $n = 8$ components, and we will use this model going forward to create our clusters labels, just like we did for $k$-means:

```
GMM = mixture.GaussianMixture(8, covariance_type='full', random_state=0)
clustersGMM = GMM.fit_predict(maskdata_scaled)
clustersGMM = clustersGMM + 1
```

The resulting predicted distribution is shown in Figure 7.23; compared to the $k$-means result, shown in Figure 7.19, we see that the GMM is able to correctly

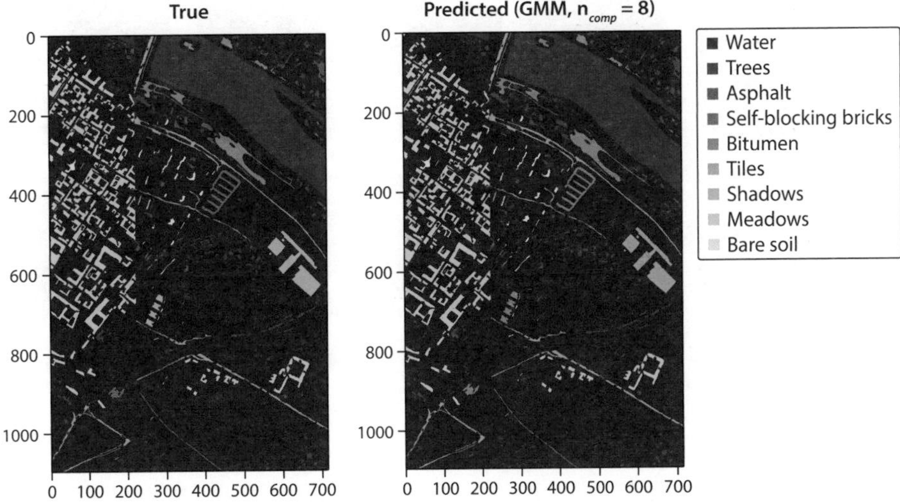


**Figure 7.24: Comparison of ground truth labels (left) vs clustering predictions after applying the centroid mapping procedure described in Section 7.5.2 (right), for an 8-component Gaussian mixture model.**

predict the presence of two dominant classes, suggesting a lower confusion. In fact, the Adjusted Rand Index for the distribution predicted by GMM and compared to the ground truth labels is $\sim 0.83$, significantly higher than the value of 0.77 found for $k$-means, even when the "true" number of classes had been set a priori.

Finally, we can re-do the centroid-mapping procedure described in the previous section to assign labels to clusters and compare the clustering result to the ground truth labels. The accuracy score achieved by GMM on nonbackground pixels is a remarkable $\sim 91\%$. We can also compare the images colored by ground truth and GMM-predicted labels to understand where errors come from; this is shown in Figure 7.24. It is quite interesting to note how, apart from lower overall error rate, the error profile of GMM is quite different from the one obtained by $k$-means: We don't observe salt-and-pepper noise (pixel-by-pixel noise), and instead see that some classes are assigned to the wrong clusters, but much more consistently (as suggested by the higher ARI score). This is easier to fix with a semi-supervised approach (by obtaining labels for some objects in a cluster), so we can conclude that the GMM method performs well in clustering these data.

## 7.6 SO CLOSE, NO MATTER HOW FAR: THE IMPORTANCE OF DISTANCE METRICS

One last concept I'd like to emphasize in this long chapter is how strongly the concepts of clustering and dimensionality reduction are connected to the concept of *distance*. Just as for supervised learning algorithms, we have learned that the

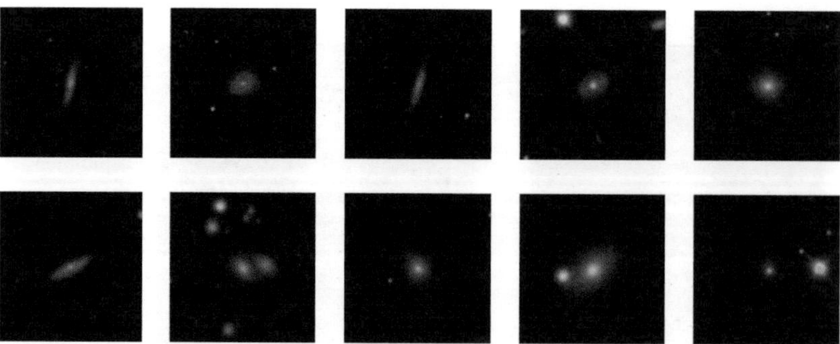

**Figure 7.25: Ten galaxies selected at random from the Kaggle Galaxy Zoo competition data set. We can visually identify at least two types of galaxies: spirals (seen at various orientations) and ellipticals.**

loss/cost function (or score) that we use affects how models are built and optimized. Here distance is crucially used to measure similarity between data points and is central to mapping procedures to new data spaces. Because of this, we need to be aware of what distance metric we are using (often, the Euclidean distance is the default choice) and ask critically whether this choice is appropriate for our data-driven problem. This can be a highly nontrivial question!

As an illustration, let's consider the example of clustering galaxies with similar morphological features. Automated classification of morphological type (e.g., spiral galaxy or elliptical galaxies) was actually one of the earliest applications of ML in astronomy, culminating with a well-known Kaggle competition won by [Dieleman et al., 2015]. In these exercises, the classes were decided a priori, and labeled examples were provided as a training set. But it's tempting to approach this as a purely unsupervised problem, and ask: *How many different types of galaxy morphologies are present in observationally complete data* (i.e., large surveys)? This is exciting, because visual inspection of galaxy images can provide a validation for usually opaque clustering results, and because we can "test" our own intuition on how many fundamental galaxy types exist, paving the way for discoveries in galaxy formation and evolution.

We use a sample data set from the Kaggle Galaxy Zoo competition,[6] and select 200 random images. After applying an admittedly rudimentary procedure (see lecture notebook "KMeans_Galaxy_Images.ipynb") to remove multiple sources from images, we can take a look at the first ten objects, shown in Figure 7.25. We can visually identify at least two types of galaxies: spirals (seen at various orientations) and ellipticals. Can a clustering algorithm do the same for us? There is only one way to find out! We can deploy our trusted friend, the $k$-means, on this data set. Note that unlike the previous example, in which each pixel was an instance and its spectrum was the set of features, here each image is an instance, and the features are

6 https://www.kaggle.com/c/galaxy-zoo-the-galaxy-challenge

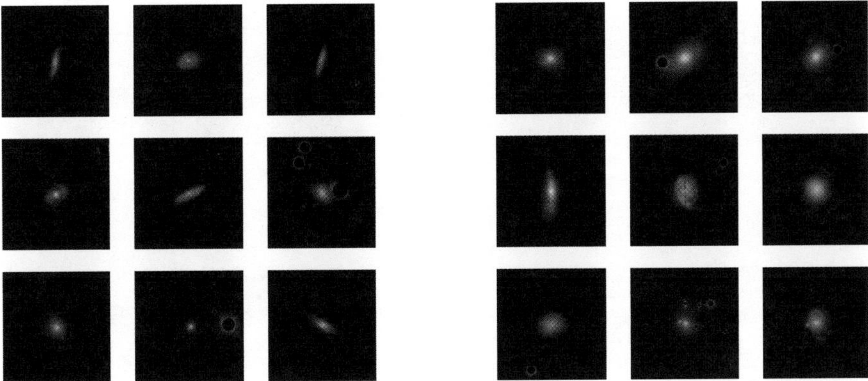

**Figure 7.26: Examples of galaxies placed by the *k*-means algorithm in opposite clusters. The main difference between objects in the two clusters (i.e., in the left and right panels) is their size, which was not the intended goal of the exercise.**

the measured intensities in each pixel. This is a fairly common setup when analyzing image data.

By running the *k*-means algorithm with two clusters, we obtain a distribution where 58 galaxies end up in one cluster, and 142 in the other one. We can then plot some examples from both clusters to see whether everything has gone according to plan; we do so in Figure 7.26. But alas, as you might have imagined from the dramatic setup, galaxies assigned to each cluster do not share the same morphology: They are instead similar in size. Why did this happen? As they say, hindsight is 20/20: Because we are using the Euclidean distance, calculated over all pixels in the image as our measure of similarity, the overall size of the galaxy dominates the similarity assessment. Two galaxies with the identical shape and morphology, but different sizes, will appear as "very different" to poor *k*-means or any other clustering algorithm based on the same distance metric.

The problem seen here is not unlike what we experienced at the beginning of our learning journey, in Chapter 2, when we first discussed the kNN algorithm and realized the importance of scaling data. One possible fix could be to normalize the galaxy size in all images, but unfortunately this is unlikely to be sufficient; there are many other potential pitfalls that we need to be careful about when choosing a distance metric. One major problem is that the Euclidean distance is not invariant under rotations: two views of the same galaxy at different orientations will not be considered "close." For this reason, it is customary to use different distance metrics when operating on images, such as the Earth Mover Distance (EMD) [Rubner et al., 2000], also known as the Wasserstein distance, or its variations. The EMD measures the amount of "work" necessary to transform one distribution into another; the name comes from the analogy with two piles of dirt (earth), where the EMD minimizes the cost of turning one pile into the other, and the cost is the amount of dirt moved multiplied by the distance moved. More recent and sophisticated

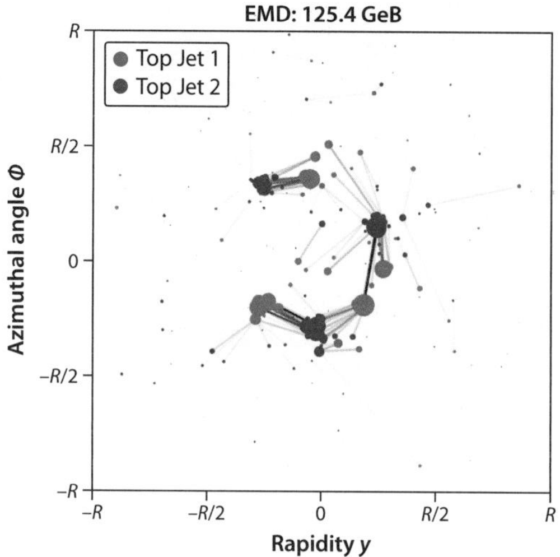


**Figure 7.27: The optimal rearrangement of one top-quark jet (red) into another (blue), in the rapidity-azimuth plane. Darker lines indicate greater movement of energy. Figure from [Komiske et al., 2019].**

approaches involve building cost functions that are invariant to specific image transformations, including not just rotations, but also blurring or scaling; see for example [Ji et al., 2019].

In Figure 7.27, we show an example from the realm of particle physics, where different jet collision events, identified by the trajectories of particles in the collider, are compared using the EMD, which measures the amount of work necessary to transform one example into the other by means of simple unit translations.

## 7.7 OTHER NONLINEAR MAPPING TECHNIQUES

Visualization of high-dimensional data sets is an important component of research in physics and astronomy, given the complex nature of the natural world. Simple techniques like linear PCAs are very useful and can be easily interpreted, but they have limitations. Many other nonlinear dimensionality reduction techniques can be used with success to detect and visualize patterns in the data besides the Kernel PCAs already mentioned; for example, see the `sklearn` User Guide[7] for a concise visual introduction.

Here we discuss briefly two of these methods, Self-Organizing Maps (SOMs) and t-distributed Stochastic Neighbor Embedding (t-SNE) ; we will consider a third one, AutoEncoders, in Chapter 8, when we discuss neural networks. See [Baron, 2019] and many of the applications described in [Feickert and Nachman, 2021] for a collection of practical examples.

7 https://scikit-learn.org/stable/modules/manifold.html

### 7.7.1 Self-organizing maps

SOMs [Kohonen, 1997] are a hybrid clustering-dimensionality reduction method that aims to find representative examples of the data, not unlike the centroids of *k*-means, and to organize them in a map so that more similar objects are closer to each other. The topology of the map is determined by the characteristics of the data.

A map is characterized by its shape: usually a $2 \times 2$ set of cells, somewhat confusingly called "neurons." Each neuron is a vector in feature space and is associated with a location in the map. In the beginning, neurons are generated at random, but they evolve during the training process. At each iteration, a sample is picked at random from the data, and the closest neuron (known as Best Matching Unit, or BMU) is selected. The coordinates of the neuron closest to the training sample and its immediate neighbors are then updated by shifting them toward the training sample by an amount that decreases as a function of distance. The BMU shifts the most, while neighboring neurons are less affected. The process is then repeated for all the data; the hope is that at the end, the neurons are able to provide a good representation of the data (i.e., all data points are close to their BMUs), and the location in the two-dimensional map can be interpreted as a measure of similarity *in the original, high-dimensional feature space*. An SOM is not unique, because the size, shape, and weight function that determines the shift of neurons are determined by the user; in a purely unsupervised approach, such parameters can't be tuned using CV. However, some measures of the quality of the SOM exist, including the *quantization error* and the *topographic error*. The former describes the average difference between data points and their BMUs: A low value points to a highly accurate representation. The latter is a measure of how much adjacency in the two-dimensional space resembles adjacency in the original high-dimensional space: A low value indicates that we can "trust" the two-dimensional arrangement to resemble the original one. The masters thesis [Yuan, 2018] is a great resource to understand SOMs more in detail; in terms of Python implementation, I am partial to the `MiniSOM`[8] library, which also contains numerous informal examples.

### 7.7.2 t-SNE

The t-distributed Stochastic Neighbor Embedding (t-SNE, [van der Maaten and Hinton, 2008]) is another popular option to visualize and explore high-dimensional data. Its goal is to create a low-dimensional representation of data that

8 https://github.com/JustGlowing/minisom

preserves the concept of similarity; ideally, objects that are similar in the high-dimensional representation will be close in the t-SNE map. The main difference with respect to PCA-type methods is that here we put the emphasis on preserving *local* distances (the "neighbor" part of t-SNE): To create the mapping, we only use information about each object's distance from its neighbors within a given radius (unlike PCAs, where maximizing the variance means that we tend to preserve *large* distances).

A t-SNE map is created as follows. First, build pairwise local measures of similarity in the high-dimensional map, and express them as a conditional probability density, $p_{ij}$, by "weighing" distance with a Gaussian function. Then create the low-dimensional map: Start by generating one map at random, calculating the local pairwise conditional probability of similarity $q_{ij}$ (using the Student-t distribution, not the Gaussian, to better accommodate large distances), and calculate the loss function of such mapping using the Kullback-Leibler (K-L) divergence, which measures the distance between probability distributions. Then use gradient descent to navigate along the negative gradient of the loss function, until the map converges, or a preset number of iterations is reached.

One important parameter of the t-SNE map is the *perplexity*, which controls the width of the Gaussian density estimate in the high-dimensional map. The actual width varies from point to point, with the goal of keeping the number of neighbors considered in the pairwise probabilities constant (to account for areas of different densities in the domain). Note that because we have an explicit loss function (the K-L divergence), even if our data are unlabeled, we can tune the perplexity by creating a t-SNE map for different values and selecting the value that leads to the lowest loss (i.e., the maximum similarity between the pairwise probabilities in the original and mapped domain). Note that because of random initial conditions and the iterative gradient descent process, a t-SNE map is not guaranteed to converge within a given number of iterations. We can use the same strategy (computing the K-L divergence for several maps and choosing the map with the lowest loss) to solve this issue, similarly to what is done in $k$-means to alleviate the problem of the dependence on the initial conditions.

t-SNE maps have been shown to be successful at capturing complex manifold structures in high-dimensional spaces, and they are a powerful tool for visualization and data exploration. Some of their limitations include interpretability (if data are unlabeled, and there isn't a clear emerging structure in the data) and time complexity (pairwise probabilities take a long time to compute for large numbers: $\mathcal{O}(N^2)$). Laurens van der Maaten's github page[9] has a great collection of useful resources on t-SNEs; they are implemented as one of the methods in the `sklearn.manifold` library.

9 https://lvdmaaten.github.io/tsne/

## 7.8 SUPERVISED OR UNSUPERVISED DIMENSIONALITY REDUCTION?

In this chapter, we have discussed a few unsupervised dimensionality reduction techniques and have learned that they are often used to explore, visualize, and understand the data, well beyond the namesake task of making the size of the data smaller.

However, making data sets more compact while minimizing loss of information is a very important and common data processing task, and it is worth summarizing different options introduced throughout this book.

We have seen at least three possible strategies for dimensionality reduction:

- Techniques like PCA project the data onto a lower-dimensional space;
- Regularization, in particular regularization techniques that favor sparsity in the coefficients, like Lasso, is effectively a strategy to reduce the number of features used and thus the dimensionality of data space;
- Feature selection methods that eliminate some features based on some importance ranking, for example those seen for tree-based methods in Chapter 6, can also be used to reduce the dimensionality of data space.

As usual, we should decide between one or another method based on what we are trying to achieve. There are important differences between unsupervised techniques, such as PCA, and supervised techniques, such as regularization or importance-based selection. First of all, the former are completely blind to the problems we are trying to solve. This is a curse and a blessing: On the plus side, we obtain a broadly valid new representation that can be used for any problem. On the minus side, the measure of distance used by PCA might not be relevant for all problems, so even if we project on a subspace with the maximum variance captured, features that are relevant for specific problems might still be washed away. We discussed this in Section 7.4.3, where we saw that while the overall shape of galaxy spectra can be represented quite well by a few PCs, narrow features such as absorption or emission lines, which have lower variance, tend to be represented less well in the new space. This issue personally cost me a lot of time as I was looking for a solution to a domain adaptation problem in [Acquaviva et al., 2020]; I concluded that looking for a less redundant representation than spectra was a worthwhile preprocessing step. It is also worth noting that PCA and other manifold embedding methods, such as t-SNE, effectively perform feature engineering, so we lose the physical meaning of original features when we work in the new space. This may be a bad thing, but also a good thing, as the mapping between the two spaces can teach us something new and be used in a creative manner. One example I loved is [Bothwell et al., 2016], in which the authors used PCA to figure out a new scaling relationship between physical variables by setting the smallest PC coefficient to zero.

The supervised dimensionality reduction methods, such as regularization or feature selection, are more limited in scope, because they require building a model first; only at that point can we use the results to determine what features are important, but the results will be valid *only for that specific problem*. The feature importance derived for a given particle collision problem won't necessarily be relevant for other problems in particle physics; the parts of the spectra that are essential for estimating the mass of a galaxy (typically, the rest-frame near-infrared) will change if we are trying to determine a different physical property, for example, stellar age. On the positive side, this means that we can minimize the decline in performance of our model, even in the lower-dimensional space; and at the same time, we keep interpretability high, because we keep working in the space of physical features and just eliminate some of them. On the flip side, we lose generality, as we would need to start from scratch again when facing a new problem. It is worth noting that, technically, doing feature selection using the entire learning set introduces leakage between train and test labels (the model picks features that give the best results on the test set). While this is unlikely to have a huge impact, a possible solution is to pick the "average" best features within a cross-validated model.

## 7.9 LESSONS LEARNED

This was another very long and broad chapter, where we hopefully learned a few things! Here is a quick list:

- We introduced several clustering algorithms, from the "classic" $k$-means, to a spectral clustering algorithm like DBSCAN, to a generative model like GMM. A brief summary of their characteristics is shown in Table 7.2.
- For each of them, we discussed strengths and weaknesses: for example, $k$-means is simple and fast, but only able to find spherical clusters, while GMM with full covariance is more flexible, but requires a large training set because of the large number of parameters involved.
- We learned that the question "how many clusters are in the data?" often has a murky answer; this is because the answer will be algorithm-dependent, and because often clusters are not clearly separated, so there simply isn't a clear-cut answer. It is normal for the answer to vary when different algorithms or criteria are used; such ambiguity does, in fact, contain information!
- We discussed several criteria to evaluate how many clusters emerge from the data, given a clustering method: the elbow method, the silhouette score, and the Bayes Information Criterion (BIC), which requires a probabilistic interpretation only afforded by generative models, such as GMM.

Table 7.2: **A summary of some characteristics of clustering algorithms we introduced in this chapter. "Hyperparams" refers to hyperparameters that need to be specified beforehand (or optimized); "Prob" indicates whether the algorithm makes a probabilistic clustering assignment; "Evaluation" refers to possible strategies to assess the success of the clustering scheme; "El" is the elbow method; "Sil" is the silhouette score; "BIC" is the Bayes Information Criterion.**

| Algorithm | Need to know $k$? | Shape | Hyperparams | Prob? | Evaluation |
|---|---|---|---|---|---|
| $k$-means | yes | globular | no | no | El, Sil |
| DBSCAN/OPTICS | no | any | $\epsilon/\xi$ | no | El, Sil |
| GMM | yes | convex | cov type | yes | BIC |

- We learned about several dimensionality reduction techniques, from linear ones like PCAs, to more complex, nonlinear ones like Kernel PCA or t-SNE. We discussed how information loss and interpretability change across the spectrum of these techniques.
- We emphasized how DR techniques are often used for visualization and data exploration, beyond their original "named" goal of actually making the data smaller.
- We compared supervised and unsupervised DR techniques; supervised ones, such as regularization or feature selection, have the curse and blessing of problem-awareness, and they don't change the feature space. Unsupervised techniques are more general but require mapping to a nonphysical space, and they might not be as effective for problems where low-variance features are important.
- We learned that the concept of distance metrics is crucial in both clustering and DR, as distance is an effective cost function for this type of algorithms; using a relevant distance metric is crucial for the success of our methods.
- Finally, we realized that the distinction between clustering/DR is blurry, and that unsupervised techniques can be used in many "creative" ways, from visualization, to data exploration, to data preprocessing and feature engineering, as generative models, and so on.

## 7.10 REVIEW AND DISCUSSION QUESTIONS

Note: Questions and exercises marked by ** are more complex, open-ended, or time consuming. Those marked by * have more than one correct answer.

**Exercise 1.** Which statement best summarizes the objective of a clustering algorithm?

A) To build clusters in which the elements of each cluster are as close as possible to one another, and different clusters are as close as possible to one another.

B) To build clusters in which the elements of each cluster are as far as possible from one another, and different clusters are as far as possible from one another.

C) To build clusters in which elements of each cluster are as far as possible from one another, and different clusters are as close as possible to one another.

D) To build clusters in which the elements of each cluster are as close as possible to one another, and different clusters are as far as possible from one another.

**Exercise 2.** The points in Figure 7.28 have been divided into the three clusters shown. This is an example of:

A) Partitional clustering
B) Hierarchical clustering
C) Fuzzy clustering
D) Partial clustering

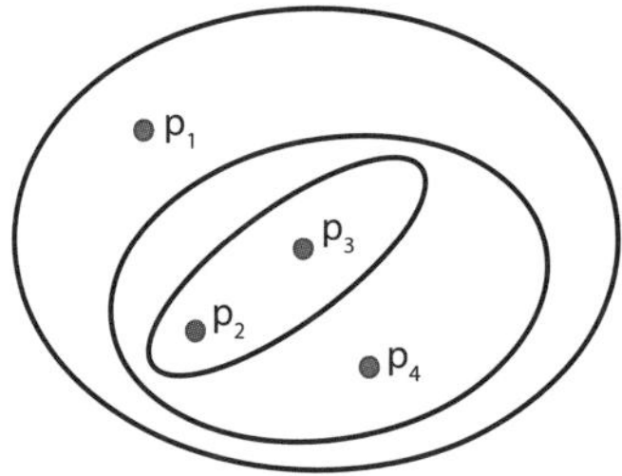


**Figure 7.28: Some points in clusters. Example from this webpage.**[10]

**Exercise 3.** * Select all true statements. Clustering algorithms can be used to. . .

A) Reduce the dimensionality of data space
B) Label unknown instances
C) Group together similar instances
D) Predict an unknown target

**Exercise 4.** Considering the $k$-means algorithm, if points $(-1, 3)$, $(-3, 1)$, and $(-2, -1)$ are the only points assigned to the first cluster now, what is the centroid for this cluster?

**Exercise 5.** * Which of the following statements about the $k$-means algorithm are correct? Select all that apply.

A) The $k$-means algorithm is sensitive to outliers.
B) For different initializations, the $k$-means algorithm will definitely give the same clustering results.
C) The centroids in the $k$-means algorithm may not be any observed data points.
D) The $k$-means algorithm can detect non-convex clusters.

**Exercise 6.** Which of these methods *cannot* be used as a measure of the goodness of clustering?

A) Elbow method
B) Clustering h-index
C) Silhouette score
D) Bayes Information Criterion

**Exercise 7.** A silhouette score value close to zero indicates that. . .

A) The sample has been assigned to the wrong cluster.
B) The sample is appropriately clustered.
C) There are overlapping clusters.

**Exercise 8.** * Gaussian mixture models (GMMs) are more powerful than $k$-means because. . . (select all that apply)

A) They allow a probabilistic interpretation of results.
B) They are faster.
C) They have fewer parameters.
D) They can be used as a generative model.
E) They can deal with nonspherical shapes.

**Exercise 9.** Which one of these methods can be used to measure how well GMM works as a density estimator/generative model?

A) Gap statistics
B) Bayesian Information Criterion
C) Silhouette score

**Exercise 10.** * Which of the following statements are correct? Select all that apply.

10 https://jcsites.juniata.edu/faculty/rhodes/ml/clusterAn.htm

A) $k$-means will succeed in clustering objects, independent of the shapes of the groups present in the data.
B) Clustering methods can give you nonsensical results if the distance metric used in clustering is not a good measure of similarity.
C) The elbow method is all about finding the point of inflection on a graph of cost function (inertia) as a function of the value of $k$.
D) GMM is good at clustering when dealing with spherical cluster shapes, but it performs poorly when dealing with elongated shapes.

## 7.11 PROGRAMMING EXERCISES

**Exercise 1.** For the data set of Section 7.4.2, figure out how many components are needed to keep all spectra below the 10% fractional difference line for linear PCA.

**Exercise 2.** ** For the same data set, use the `KernelPCA` library in `sklearn` to set up a dimensionality reduction pipeline analogous to the one we built for linear PCA, using the Gaussian kernel. Include a hyperparameter optimization search for $\gamma$. Decide on a measure of success for your algorithm. For 10 components, are the results better or worse than what we obtained for linear PCA? How about for 50 components? What can you conclude about the better DR method for this problem?

**Exercise 3.** ** In the remote sensing application, discuss what happens if one applies the dimensionality reduction before or after the clustering, for $k$-means or GMMs.

**Exercise 4.** ** Build a two-dimensional map of the Pavia remote sensing data set (or another classic hyperspectral images data set; many of them are available here),[11] using a SOM or t-SNE. Use the evaluation tools described in this chapter, for example, quantization/tomographic error for SOMs, and K-L divergence for t-SNE, to evaluate the map. Then "color" the map according to the ground truth labels. Comment on the results.

11 http://www.ehu.eus/ccwintco/index.php/Hyperspectral_Remote_Sensing_Scenes

CHAPTER EIGHT

# 8 Introduction to Neural Networks

*What does the successful neural network say to the unsuccessful neural network?*
*I am sorry for your loss.*

This chapter is a very limited introduction to the world of deep learning, which offers some of the most powerful and flexible ML algorithms: neural networks. Deep learning is a subdiscipline of machine learning that is concerned with building, training, and optimizing neural networks.

## 8.1 DEEP LEARNING AND WHY IT WORKS

What makes a deep learning algorithm? I would say that the key issue is to have several *layers* of input/output relationships. In all the algorithms we have seen so far, there was one implicit function, however complicated, that related the input features to the output (although one could possibly argue that kernel methods such as SVM also have a built-in feature engineering step). In deep learning algorithms, we "stack" several of these functions, so that the outputs of one layer become the inputs of the next one. The parameters of the algorithm are distributed along all possible connections, and they measure the strength and the functional form of each connection. The scheme of the connections between the inputs and outputs forms a network, and the "*neural*" attribute comes from the resemblance between this scheme and the network of synapses in the human brain.

The concept of neural networks has been around for many decades [McCulloch and Pitts, 1943], but their popularity has exploded in the past decade or so, with immense progress being made in a short amount of time. Currently, they have grown to dominate the landscape of ML-based papers in physics and astronomy.

The reasons for their recent increased popularity have a lot to do with the increase in data set size and complexity, but also, in my opinion, with accessibility.

Neural networks are considerably more complicated than any other method to code from scratch, and they require significantly higher computing power (e.g., graphics processing units) to train, because they tend to have many parameters, easily on the order of thousands or millions, that require optimization. Until recently, the lack of access to suitable computing platforms and resources was an issue; this has been significantly alleviated by the advent of open-source frameworks like `TensorFlow` [Abadi et al., 2015] and `Pytorch` [Paszke et al., 2019], by high-level application programming interfaces such as `keras` [Chollet et al., 2015] that further simplify the process of building and training deep learning models, and by the increased access to graphics processing units and tensor processing units (see also Section 9.2.1).

We are fortunate to be able to unleash the power of neural networks easily and conveniently, provided that we understand what we are doing. In this chapter, I provide a simple introduction to the building blocks of neural networks; many excellent resources are available for those who are ready to continue their deep learning journey as practitioners.

Before moving on to the computational aspects of deep learning, we can take a look at the strengths and weaknesses of neural networks. What makes them so powerful? Their number one strength, at least in my opinion, is *flexibility*. There is a lot of freedom in choosing a neural network's architecture and customizing its elements, from the different layer structures to the loss function. As a result, they can be used to successfully solve a variety of classification and regression problems, and they are well suited to learn from large and complex data sets. The flip side of flexibility is, unsurprisingly, *complexity*: both computational complexity, deriving from the large number of parameters that make neural networks slow to train, and interpretation complexity, the fact that the decision-making process of a neural network is considerably more opaque than that of simpler algorithms.

In the following section, we will start assembling simple neural networks, piece by piece.

## 8.2 ASSEMBLING A NEURAL NETWORK

The simplest neural network that can be built consists of two elements: a linear model and a nonlinear function that processes the output of the linear model. If this nonlinear function is the Heaviside step function (a function that is 0 when $x < 0$, and 1 otherwise; Figure 8.7), our tiny machine is called a *Perceptron*. Figure 8.1 shows an example where the model has three input features, and one more (the bias) is added with a constant value of 1, as shown previously in Chapter 5, so that we can write the output of the linear model in matrix multiplication form: $y_{\mathrm{LM}} = \mathbf{x} \cdot \mathbf{w} = w_0 + w_1x_1 + w_2x_2 + w_3x_3$. The output of the linear model is then processed through the nonlinear function, in this case the Heaviside function,

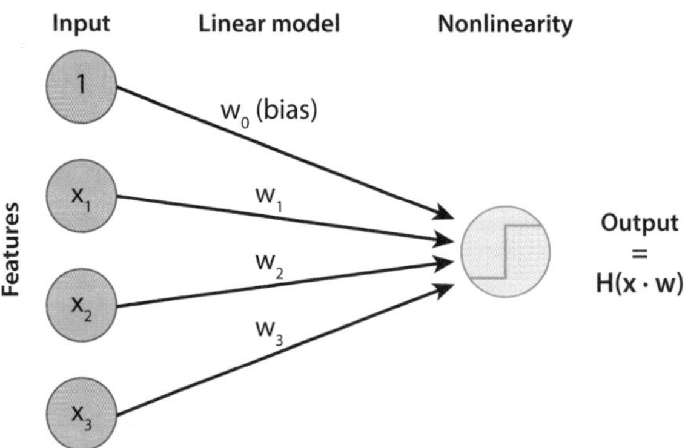


Figure 8.1: A tiny, single-layer Perceptron.

so that the final output of this layer is $H(y) = H(\mathbf{x} \cdot \mathbf{w}) = H(w_0 + w_1x_1 + w_2x_2 + w_3x_3)$.

This model has a total of four parameters: the *bias* $w_0$, and the *weights* $w_1, w_2$, and $w_3$, which measure the strength of the association between the three features and the final output.

Is the Perceptron a neural network? We could say so, because it involves two different stages: the linear model and the nonlinear function. However, this sounds mostly like a semantic issue, and we can probably agree that not many interesting problems can be solved by just using the Perceptron. In fact, scientists abandoned the idea pretty quickly in the 1960s, because the Perceptron could not even solve simple logical problems, such as the XOR problem (see Section 8.2.1). Chapter 10 of [Géron, 2019] contains a fun recollection of the Perceptron's historical demise. As often happens in life as in machine learning, the disappointment was mostly a matter of misguided expectations. The Perceptron is merely a linear classifier in disguise: All it can do is to draw hyperplanes and then assign objects to one class or another, according to which side of the hyperplane they lie on.

### 8.2.1 Multi-Layer Perceptron

What the single Perceptron cannot achieve, a team of Perceptrons can do! Things quickly become more interesting if one stacks together a few Perceptrons to obtain a Multi-Layer Perceptron, or MLP. In two dimensions, it is easy to visualize the capabilities of a series of connected linear classifiers: We can draw one dividing line, and then another, and then another, and at each step, we assign objects on each side of the line to our class of choice.

Let us apply this newfound knowledge to solving the XOR problem. It is a two-coordinate problem in which the inputs (0,0), (0,1), (1,0), and (1,1) need to return outputs 0, 1, 1, and 0. Figure 8.2 represents the outputs 0 and 1 by circles

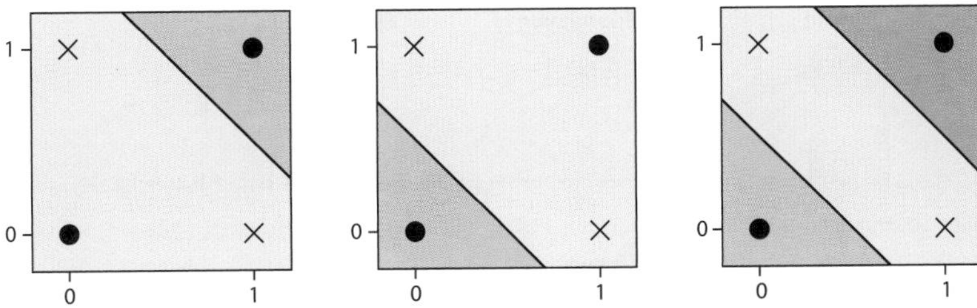


**Figure 8.2: Diagrams for single Perceptrons (left and middle panels) and their combination, corresponding to an MLP that can solve the XOR problem (right panel). In each diagram, the light green region shows the points that are mapped to 1; the light blue region shows the points that are mapped to 0. The four symbols show the "true" labels of the XOR problem; filled circles are 0 (False) and crosses are 1 (True).**

and crosses, respectively. While one single Perceptron is not able to solve this non-linearly separable problem, it is clear that two lines, drawn for example at a $45^\circ$ or $-45^\circ$ angle, would be sufficient. For example, we could use one Perceptron to draw the line and create the classes (represented by background colors) of the left panel of Figure 8.2, and another Perceptron to draw the line and create the classes of the middle panel. In the first diagram, points to the right of the dividing line (in the light blue area) are classified as a 0, and points to the left of the dividing line (in the light green area) are classified as a 1. In the middle diagram, the decision is reversed: Points to the right of the dividing line (in the light green area) are classified as a 1, and points to the left of the dividing line (in the light blue area) are classified as a 0. The final step is combining the two Perceptrons: In this case, we can just take their outputs, add them together, and then "pass them" through the Heaviside function again (which, by the way, doesn't alter the results). The outcome of the combined Perceptrons is shown in the right diagram: Points in the bottom left and top right (light blue area) are classified as a 0, while points in the region between the two lines (light green area) would be classified as a 1. This diagram correctly solves the XOR problem.

We can, in fact, draw the scheme and figure out the weights and biases of the tiny neural network corresponding to the diagrams above.

The first step is to figure out the architecture. We know already that we have two input features (the x and y coordinates, denoted by $x_1$ and $x_2$), and that we need to stack two Perceptrons: We will have one input layer, one intermediate layer (corresponding to the output of the first Perceptron, and the input of the second), and one output layer. The intermediate layer is a *hidden layer*; this term applies to any layer between the input features and the final output.

The diagram of our MLP is shown in Figure 8.3. To build an actual model, we need to figure out suitable values for the weights. We can start by counting them, layer by layer. If $M$ is the number of input features, and $N$ is the size of the intermediate layer, the first layer has $N + 1$ neurons (because of the bias), and each of

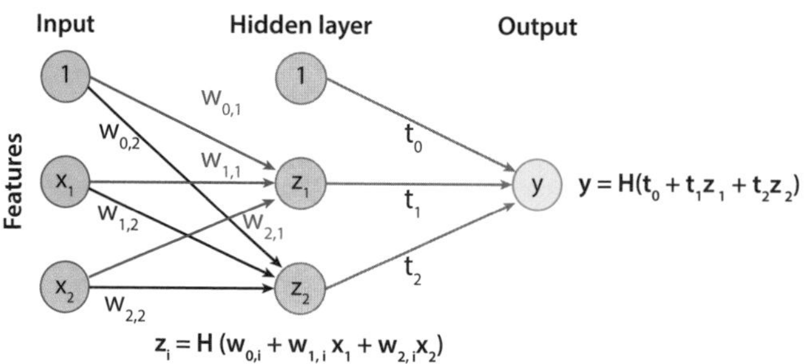


**Figure 8.3: Diagram of an MLP with one hidden layer that can solve the XOR problem.**

them is connected to the $N$ neurons of the next layer, for a total of $(M+1) \times N$ weights. In this case, $M=N=2$, so we have six weights. In the next step, we have $N+1$ neurons in the intermediate layer, and each of them is connected to only one neuron in the output layer. Therefore, this step adds $(N+1) \times 1=3$ weights, for a total of nine weights in the MLP. Networks like this one, where each neuron talks to all the neurons in the preceding and following layer, are called *fully connected* neural networks. In such systems, if $k$ is the total number of layers, the total number of weights will be

$$N_w = \sum_{0}^{k-1} (N_i + 1) \times N_{i+1}, \tag{8.1}$$

where $N_i$ is the number of neurons in the $i$-th layer.

We can finally take on the task of figuring out the actual values of all the weights. There are many possible solutions, but let us focus on the one exemplified by the diagrams of Figure 8.2. The set of connections between the input features and the first intermediate neuron, $z_1$, represents the split of the left diagram. The equation of the line, in slope-intercept form, is $x_2 = -x_1 + 1.5$ (the $x$, $y$ coordinates are named $x_1, x_2$ to avoid confusion with the output of the MLP). We can write it equivalently in "boundary" form: $x_1 + x_2 - 1.5 = 0$. From this, it would be tempting to read the weights right away. However, note that using the weights from this equation would give the opposite diagram, mapping points to the right of the line to 1, while they should be mapped to 0. Therefore, we need to take the weights with a negative sign in order to represent the diagram correctly: In the notation of Figure 8.3, $w_{0,1} = 1.5$, $w_{1,1} = -1$, and $w_{1,2} = -1$. The output of the first neuron in the intermediate layer will be $z_1 = H(-x_1 - x_2 + 1.5)$, and you should convince yourself that it correctly represents the results of the left diagram; for example, the points (0,0), (0,1), and (1,0) are mapped to 1s, while the point (1,1) is mapped to 0.

Now consider the set of connections between the input features and the second intermediate neuron, $z_2$, which should represent the split of the middle diagram in Figure 8.2. Again, we start by writing the equation of the dividing line in slope-intercept form: $x_2 = -x_1 + 0.5$, which can be rewritten as $x_1 + x_2 - 0.5 = 0$ to obtain the weights. In this case, the separation induced by the Heaviside function on this line correctly matches the diagram, because points to the right of the line are mapped to 1; the weights are $w_{0,2} = -0.5$, $w_{1,2} = 1$, and $w_{2,2} = 1$. The output of the second neuron in the intermediate layer is then $z_2 = H(x_1 + x_2 - 0.5)$, which maps the point $(0,0)$ to 0, and the points $(0,1)$, $(1,0)$, and $(1,1)$ to 1, matching the middle panel. The final step is to find the weights for the connections between the hidden layer and the final output. If we simply add the outputs of the two single Perceptrons, we obtain two regions matching those of the right panel that we want to represent, but the output values would be 1 in the outer regions, and 2 in between the lines. If passed through the Heaviside functions, all those points would be mapped to 1s! Thus we need to use our bias neuron (the $t_0$ weight) to shift the output, for example by $-1.5$ (but many values are possible, as long as they map the two classes to a negative/positive value, so that the Heaviside function can do its job)! In the notation of Figure 8.3, $t_0 = -1.5$, $t_1 = 1$, and $t_2 = 1$. The final output of the network is $y = H(-1.5 + z_1 + z_2) = H(H(-x_1 - x_2 + 1.5) + H(x_1 + x_2 - 0.5))$.

### 8.2.2 Activation functions

The discussion in the previous section offers a glimpse of why stacking multiple units, where a unit is considered a combination of a linear model plus a nonlinear *activation function*, improves the capabilities of deep learning models. More generally, we can define a layer of a neural network as the combination of two steps: (1) performing a linear operation on the input features, and (2) "passing the result" through a nonlinear activation function.

Activation functions play an important role in neural networks, and recent research has shown vast performance improvements as a result of "smarter" activation functions, as discussed in Section 8.3.2. But why do we need activation functions in first place? The simple answer is that *without nonlinear activations, fully connected neural networks are just linear models.* You can convince yourself that this is the case by looking at the output of the MLP from the previous section: $y = H(-1.5 + z_1 + z_2) = H(H(t_0 + t_1\, H(w_{0,1} + w_{1,1}x_1 + w_{2,1}x_2)) + t_2\, H(w_{0,2} + w_{1,2}x_1 + w_{2,2}x_2))$. If we remove the Heaviside function from this expression, we obtain a linear combination of the original features plus a bias term: $y = t_0 + t_1\, w_{0,1} + t_2\, w_{0,2} + (t_1\, w_{1,1} + t_2\, w_{1,2})\, x_1 + (t_1\, w_{2,1} + t_2\, w_{2,2})\, x_2$.

Besides the essential requisite of nonlinearity, other characteristics, such as differentiability and symmetry around 0, are desirable in activation functions. This is why the Heaviside function, which has neither, was never a popular choice of activation function, and it was quickly replaced by the *sigmoid* activation, already encountered in our discussion of logistic regression in Chapter 5, or the *hyperbolic tangent* activation (*tanh*).

These functions are more suitable choices, because the process of training a neural network, just like any ML algorithm, requires optimization, and the optimization passes through the minimization of a global loss function. You won't be surprised to encounter here our old friend, gradient descent! A good activation function will *preserve* the gradients of the network, making them easy to compute, not too large, and not too small. We will look at how these good qualities can be enabled in the following sections.

### 8.2.3 The Universal Approximation Theorem

Just as for LEGOs, once we have the fundamental building blocks, a.k.a. layers, we can build neural networks of arbitrary complexity. We have freedom in choosing the size of the hidden layers, which is the *width* of the network, and number of layers to stack, which is the *depth* of the network. We can also customize the activation functions chosen for each layer. One final word of wisdom about the activation of the final layer: If we are dealing with a classification problem, then activations like the Heaviside function, the sigmoid function, or the hyperbolic tangent function will work, but a regression problem will usually just have a linear function in its final layer. This is OK as long as we have nonlinear activation into the previous layers, so we don't fall into the trap of the fancy linear model.

By increasing the width and depth of a neural network, we are able to represent increasingly complex input/output relationships. It turns out that neural networks of arbitrary size can represent *any* function: This is known as the *Universal Approximation Theorem* (see, e.g., [Cybenko, 1989; Hornik et al., 1989]). Other than being properly impressed, we won't consider the theorem in much detail here, but you can enjoy a somewhat intuitive, interactive explanation of it in Chapter 4 of [Nielsen, 2018]. This theorem is the formal seal of the acclaimed flexibility of neural networks as ML models.

## 8.3 HAVE NETWORK, WILL TRAIN

Once we have built an architecture for our (still fully connected) neural network by stacking together some layers of different sizes, the big (and time-consuming) issue is: How do we train it?

First of all, we should clarify what it means to "train" a neural network of a given shape: It means that *we need to find the values of weights and biases that minimize a given loss function*. Defining a loss function is nothing new: We are already familiar with many of them, for example, the mean square error (MSE) for regression problems and the cross entropy for classification problems.

In the XOR example discussed in Section 8.2.1, we found the weights of the network by hand, and to be fair, we just showed that they correctly solved the XOR problem, but we didn't ensure optimality in any way, and we defined our loss function only implicitly as the accuracy on the four samples with coordinates (0,0), (0,1), (1,0), and (1,1). In the general case, we will need an optimization machinery.

The process of training a neural network is an iterative process that works as follows:

1. First, initialize the weights and biases.
2. Next, for every example in the training set, calculate the loss function. This step is known as the *forward pass*, because we start with the input features in the first layer, and propagate those through the network until the final (output) layer. This is a common step to all ML models; we are just doing it multiple times, because we have multiple layers. The outcome of this step is some value of the loss function, which probably won't be the minimum.
3. The next step consists of adjusting the weights in order to move toward the minimum of the loss function. In principle, this is nothing new, just another day in gradient descent world. However, the procedure of calculating the gradients (which are the derivatives of the loss function with respect to all the weights and biases) for a neural network is nontrivial.
4. Now take a gradient descent step, adjusting all the weights and biases in order to move to a lower value of the loss function. As a side note, it is common to perform steps 2–5 using either stochastic gradient descent or mini-batch gradient descent, rather than going through the full data set at once. The process of calculating the gradients efficiently and "distributing" them along the network to take a gradient descent step is called *backpropagation*, and the next subsection is dedicated to it. For now, let's just assume that we have the gradients ready.
5. Repeat the loop of forward pass/backpropagation until satisfied with the loss function value, or more commonly, for a given number of full training loops, called *epochs*.

We will play with training a fully connected neural network soon, but first, let's take a closer look at how backpropagation works.

### 8.3.1 Backpropagation

Backpropagation is the cornerstone of the training process for deep neural networks. In this step, we determine how the different connections of the network contribute to the loss function and use this information to optimize the strength of the connections, which are the weights.

We can understand what needs to happen in graphical form by looking at Figure 8.4, which depicts a simple network with three layers before the output. Each arrow indicates a connection (or edge). To understand how gradients transfer across layers, we can start by drawing a simple *derivative graph*, and ask: What is the derivative

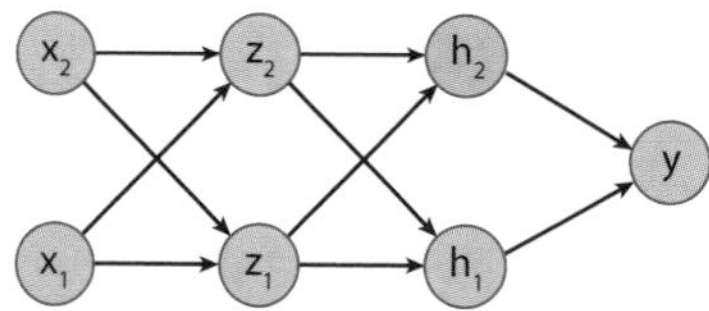


**Figure 8.4: A graph showing a layered input-output structure similar to a neural network (without bias neurons).**

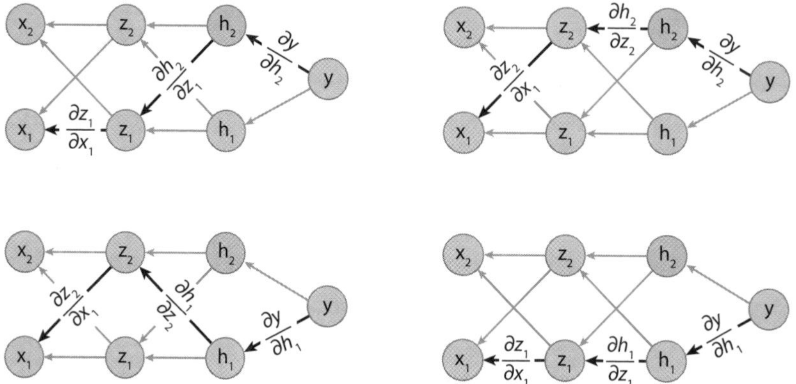

**Figure 8.5: Computational derivative graph, showing the four paths that lead from *y* to $x_1$. Each component is obtained by multiplying the three partial derivatives on each path, and the four paths need to be added in order to calculate the derivative $dy/dx_1$.**

of the output ($y$) with respect to the first input ($x_1$)? The graph shows that the gradient $\frac{\partial y}{\partial x_1}$ can be obtained as the sum of the gradients along each path that takes us from $y$ to $x_1$, and each of the partial components is the product of the partial derivative along the edges encountered. For this particular case, the four paths are the ones shown in Figure 8.5, where the relevant partial derivative is indicated for each edge. Putting the four of them together, we obtain:

$$\frac{\partial y}{\partial x_1} = \frac{\partial y}{\partial h_2}\frac{\partial h_2}{\partial z_1}\frac{\partial z_1}{\partial x_1} + \frac{\partial y}{\partial h_2}\frac{\partial h_2}{\partial z_2}\frac{\partial z_2}{\partial x_1} + \frac{\partial y}{\partial h_1}\frac{\partial h_1}{\partial z_2}\frac{\partial z_2}{\partial x_1} + \frac{\partial y}{\partial h_1}\frac{\partial h_1}{\partial z_1}\frac{\partial z_1}{\partial x_1}. \tag{8.2}$$

This is, of course, not an innovative result, but merely the application of the derivative *chain rule*, which tells us how to decompose gradients along each component. The derivative graph is just a useful visual tool to understand how each component corresponds to a "path" in the graph.

The task of computing the derivative of the loss function with respect to the parameters (i.e., the weights) of the network, a.k.a. backpropagation, closely resembles the scheme of our derivative graph; we will be able to recycle some of the intuition we have built for that purpose.

A complete neural network diagram would need to include the loss function $L$, which will be a function of the predictions $\hat{y}$, and of the true values $y_T$; we will also

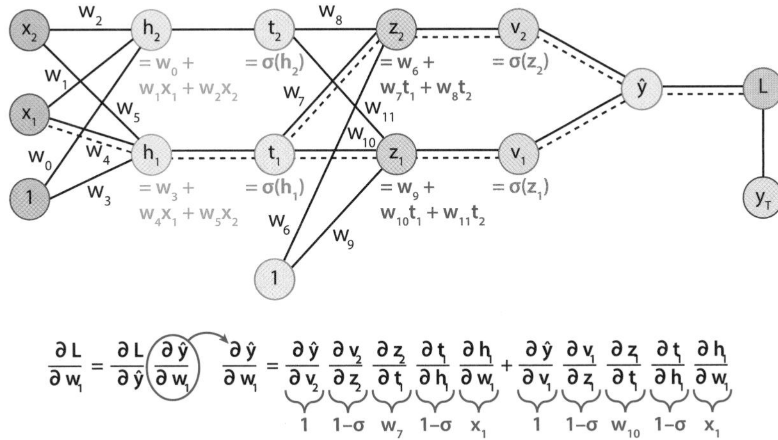

Figure 8.6: Scheme for backpropagation.

show the outputs of each neuron more explicitly and include "horizontal" steps for activation functions, $A$. Finally, keep in mind that the derivatives will be calculated with respect to the weights of the network, not the input of the neurons. An example graph for backpropagation is shown in Figure 8.6.

Let us go through the exercise of calculating the derivative of the loss function $L$ with respect to the weight $w_1$; other derivations would be very similar.

Just as for the computational derivative graph, the derivative $\frac{\partial L}{\partial w_1}$ can be obtained by multiplying all the partial derivatives found along the paths that lead from $L$ to $w_1$ and adding the contributions of all the possible paths. This is, of course, just a visual way of applying the derivative chain rule. In this case,

$$\frac{\partial L}{\partial w_1} = \frac{\partial L}{\partial \hat{y}}\frac{\partial \hat{y}}{\partial w_1} = \frac{\partial \hat{y}}{\partial v_2}\frac{\partial v_2}{\partial z_2}\frac{\partial z_2}{\partial t_1}\frac{\partial t_1}{\partial h_1}\frac{\partial h_1}{\partial w_1} + \frac{\partial \hat{y}}{\partial v_1}\frac{\partial v_1}{\partial z_1}\frac{\partial z_1}{\partial t_1}\frac{\partial t_1}{\partial h_1}\frac{\partial h_1}{\partial w_1}, \tag{8.3}$$

and the relevant derivatives are as follows:

$$\begin{aligned}
&\frac{\partial \hat{y}}{\partial v_1} = 1,\\
&\frac{\partial v_2}{\partial z_2} = \frac{\partial A(z_2)}{\partial (z_2)},\\
&\frac{\partial v_1}{\partial z_1} = \frac{\partial A(z_1)}{\partial (z_1)},\\
&\frac{\partial z_2}{\partial t_1} = w_7,\\
&\frac{\partial z_1}{\partial t_1} = w_{10},
\end{aligned} \tag{8.4}$$

$$\frac{\partial t_1}{\partial h_1} = \frac{\partial A(h_1)}{\partial (h_1)},$$
$$\frac{\partial h_1}{\partial w_1} = x_1.$$

After this rather tedious exercise, we can cheer ourselves up by making a few observations:

- The derivatives of activation functions are prominently featured in the backpropagation algorithm, and having a differentiable function is desirable. This is one of the reasons the sigmoid function, $1/(1+e^{-x})$, immediately became more popular than the Heaviside function as an activation; $\sigma'(x) = 1 - \sigma(x)$.
- Many partial derivative terms are re-utilized multiple times; this is especially true for those in layers "closer" to the final output (to the right of the diagram), which participate in any partial derivative with respect to weights to their left. Thus a good backpropagation algorithm would find a way to optimize the gradient calculations so that we avoid recalculating those terms numerous times. In practice, this is achieved by an algorithm called *automatic differentiation,* often abbreviated as *autodiff* (or in this case, *reverse autodiff,* because we are moving backward, from output to input).
- The chain rule is the workhorse of backpropagation. In fact, a rite of passage for ML practitioners is being able to say: "Oh, backpropagation? It's really just the chain rule," in a slightly condescending tone. Congratulations, you are now part of this crowd (but please be kind and not condescending toward those who are still on the other side)!

### 8.3.2 Weight initialization and activation functions

Before seeing neural networks in action, let us consider some well-known issues that might prevent a network from functioning properly. The first one is, in fact, related to step 1 of the network training procedure highlighted at the end of Section 8.3.

How should we initialize the weights of the network? This seemingly innocuous question is actually quite nontrivial to answer, and proper initialization plays a crucial role in the proper functioning of a network. We can make some observations:

- The weights can't be initialized to the same value. If we do so, any forward pass in the network (which is entirely deterministic) creates exactly the same output for all the neurons, and the gradients stay the same throughout the training process. *Breaking the symmetry* of the network is essential. This can be done, for example, by initializing weights using a random uniform or Gaussian distribution.

- Activation functions play a significant role in how the gradients are propagated and "distributed" across layers. For example, activation functions like the sigmoid (or the hyperbolic tangent, another popular choice because of its symmetry around 0) are very flat when the inputs are large (positive or negative): In this regime, their derivative is close to 0. When we use such a derivative in the formula above, it multiplies other terms, and this drives the global gradients ($\frac{\partial L}{\partial w}$) toward very small values, preventing the network from updating the weights. This is known as the *vanishing gradients* problem.
- A possible solution for the vanishing gradients problem is to use activation functions that don't saturate for large inputs, for example, the popular ReLU (Rectified Linear Unit), which is simply defined as $max(0, x)$, is very fast to compute, and allows for sparse weight representation (i.e., the network is allowed to set weights to exactly 0). However, this same property may hinder the training: Some connections may "die," and there is no way to rescue them! The issue can be improved by using the so-called Leaky ReLU activation, which modifies the ReLU by adding a small slope (0.01) for negative inputs. Finally, the suboptimal issue of the nondifferentiability of the ReLU or Leaky ReLU activations can be improved by using a continuous and differentiable version known as ELU, or Exponential Linear Unit.
- While nonsaturating activation functions are effective at solving the vanishing gradients problem, their vigorous gradient flow might induce the opposite issue, known as the *exploding gradients* problem. A more careful approach to the issue of weight initialization is to recognize that the distribution and size of initial weights need to be adjusted in concert with the chosen activation function and with the size of the network. This approach was first explored by [LeCun et al., 1998], and more recently in [Glorot and Bengio, 2010], who framed the task of efficient training as a process that conserves information flow (i.e., the variance of inputs and gradients) as they pass through the network, in both directions. These authors derived effective rules for the initial variance of weights as a function of chosen activation (sigmoid and tanh), functional distribution of weights (uniform or Gaussian), and size of network. Their work was expanded by [He et al., 2015] for the ReLU and Leaky ReLU activations. Today the Xavier (for the first name of Glorot)/Glorot and He initializations are standard choices in neural network frameworks, such as `keras`. Figure 8.7 summarizes possible activation functions and weight optimizations.

### 8.3.3 Optimization

Just like any other ML algorithm, for neural networks, we need to optimize not only model parameters (specifically, the weights and biases), but also *hyperparameters*. The most common hyperparameters pertain to the architecture of the network, for

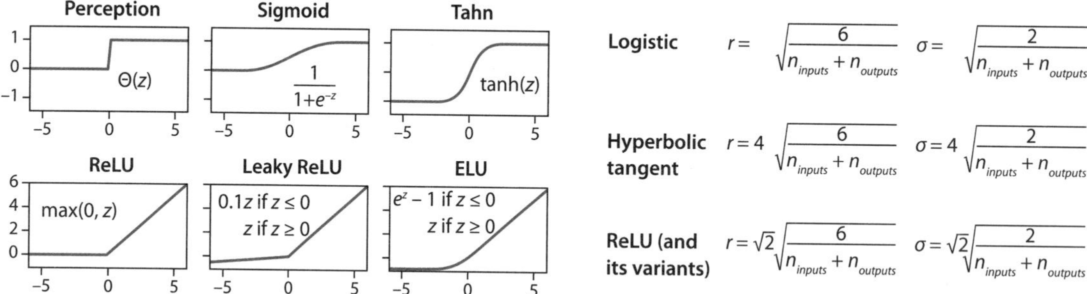


**Figure 8.7: (Left) Some common activation functions. (Right) Ad hoc initialization strategy, as suggested by [Glorot and Bengio, 2010] and [He et al., 2015], for different activation functions. The terms $n_{\text{in}}$ and $n_{\text{out}}$ are the number of input/output neurons for each layer. Inspired by [Géron, 2019].**

example, the size of each layer (which can be different for every layer!) and the width of the network (i.e., the number of hidden layers). Additional degrees of freedom will occur once we consider more specialized architectures that go beyond fully connected networks, for example, Convolutional Neural Networks (CNNs) or Recurrent Neural Networks (RNNs). Finally, the choice of activation function, weight initialization, the number of epochs, and the learning rate in the gradient descent algorithm all contribute to the success of a deep learning algorithm and also should be optimized.

It might seem like a pretty discouraging task. The good news is that we know how to do this already: Optimization can be carried out through *k*-fold CV, and remembering that different parameters are correlated, so they should be varied simultaneously, this may happen through a grid search. If, besides optimization, we also want to quote realistic generalization errors, we must use nested CV to evaluate the test scores on test folds that have not participated in the optimization process. The bad news, as you might have anticipated, is that neural networks are complex objects, and the process of training them is computationally expensive—much more so when we need to explore many hyperparameter combinations several times, as required by CV. What is a poor ML practitioner to do?

I don't have a recipe beyond common sense. In many cases, our optimization strategy will be constrained by the available computational resources, so there is no universal best. I will just list some practices that I find useful:

- Think about the data before everything else. Do you have a good representation? Does your network need to do feature engineering for you? What kind of architecture best suits your problem? In many cases, choosing the right type of network (more on this later in this chapter) is more helpful than extensive optimization.
- If your data set is large, you can select a random subset to run the hyperparameter optimization on, and use learning curves (or just compare the scores

on the full and reduced-size data set in a statistically sound manner) to make sure your results aren't affected. The final training will, of course, happen on the whole data set.

- If your resources are limited, start small and then increase as needed. Start with a few layers and a modest number of neurons.
- Focus on the hyperparameters that have the most impact. This could mean experimenting with a few values for each hyperparameter, so you can avoid optimizing those that aren't helpful.
- Build some intuition on how different hyperparameters interact; this will help determine whether some parameters can be optimized individually.
- Sample the space of hyperparameters at random instead of trying out all possible combinations, similarly to what `sklearn`'s `RandomSearchCV` utility does (compared with `GridSearchCV`); note, however, that these tools are not directly available for neural networks in `keras`.

One important aspect of optimization consists of choosing and tuning the algorithm to run the loss minimization. Chapter 5 introduced stochastic gradient descent and mini-batch gradient descent as faster alternatives to "standard" gradient descent. Here it is even more crucial to use an efficient minimization method, because the dimension of the parameter space (the number of weights/biases in the network) is very large. One problem of gradient descent is that the step size is proportional to the gradient, so it becomes inefficient (uses excessively small steps) when approaching the minimum of the loss function. This aspect can be improved through the *momentum trick*, where the correction to the weights is proportional to the accumulated *momentum*, the change in gradient over the last $N$ steps, with a small correction that depends on the gradient itself. Another solution could be to employ an adaptive *learning rate*, in which the learning rate is adjusted for different gradient directions and slopes. Algorithms such as AdaGrad [Duchi et al., 2011] or RMSprop [Hinton et al., 2020] use this approach. Currently, the most popular optimization engine is ADAM (from ADAptive Moment Estimation) [Kingma and Ba, 2014], which combines both strategies, is efficient and alleviates the problem of optimizing the learning rate, which is highly desirable. If you are curious about different optimization algorithms, you can find a very clear discussion of the pros and cons of different strategies in Chapter 11 of [Géron, 2019].

### 8.3.4 Regularization

The last trick up our beginner-neural-networkers sleeve will be regularization. You have probably guessed that because of the large number of parameters, overfitting is a real risk for deep learning algorithms! Thankfully, we have several simple strategies to choose from when the usual train/test score comparison suggests a large gap, indicative of overfitting.

1. The first strategy is known as *early stopping*. The idea behind it is simple: After the minimum of the loss function has been found, the network does not benefit from further learning. Therefore, rather than completing the planned number of training epochs, it may be beneficial to stop the learning process if the loss function starts increasing again (or if it increases for some time, if one is using stochastic or mini-batch gradient descent, whose trajectories are less smooth).
2. The second option is to use regularization, which was introduced in Chapter 5 when discussing the Lasso and Ridge regression models. The idea is the same: Add a term to the loss function that is proportional to the sum of absolute values of weights (for L1 regularization or Lasso) or the sum of their squares (for L2 regularization or Ridge). This helps reduce the size of the weights, preventing the network from amplifying the contribution of a few neurons, or setting some weights to 0 in the case of L1 regularization.
3. The last regularization strategy is as simple as it is effective, and is known as *dropout*. It consists of "muting" a fraction of the total connections (typically 20–50%) at random at each training epoch. This increases the robustness of the learned model, much like it did when we discussed tree randomization for bagging methods in Chapter 6. In `keras`, a "Dropout" layer is available. If this strategy is used, the dropout fraction becomes one of the hyperparameters that we need to optimize.

### 8.3.5 Recap: Training a neural network

The complex process of training neural networks deserves a quick summary of steps:

- Preprocess data. At the very minimum, data should be normalized, as we do every time that different features interact with one another. The building blocks of neural networks are linear models, so the same rules apply. In fact, it is sometimes beneficial to normalize the intermediate outputs as well, as they are inputs to subsequent layers. We won't discuss this here, but it's worth reading more about this process, called *batch normalization* [Ioffe and Szegedy, 2015].
- Start with a nominal configuration. Decide on architecture, number and size of layers, loss function, optimizer, learning rate, and number of epochs. My advice is to start simple and increase complexity as needed.
- Train the network and look at learning curves. Are we in a high variance or high bias regime? Reflect on which directions are worth exploring when optimizing hyperparameters. For example, in a high-bias network, we might consider adding layers, or increasing their size (the question of whether deeper or wider networks are more efficient for the same number of parameters is a subject of active debate; see, e.g., [Wu et al., 2019] and references therein), or training for a longer time. In a network that struggles with generalization, we

might add a dropout layer, consider higher dropout fractions, or employ other regularization techniques.

- Optimize the hyperparameters of the network, using best judgment in order to reduce the search space. For example, we can optimize some parameters on their own or in a subspace if we don't believe they are strongly correlated with others, use a randomized search, and use the learning curve to select the optimal range for each parameter. Select optimal parameters using $k$-fold CV. If we need to report generalization error, we need to use nested CV.
- Retrain the network on all data to build the final model.

## 8.4 TWO WORKED EXAMPLES: PARTICLE CLASSIFICATION AND PHOTOMETRIC REDSHIFTS

We can experiment with simple neural networks by "recycling" two problems presented earlier in this book: the particle classification problem of Chapter 4, where we aimed to distinguish between the production of 4 top quarks and 2 top quarks, and the photometric redshift prediction of Chapter 6. In our examples, we use `keras` with the `tensorflow` backend, but this is only one of the possible choices of platforms.

Importing packages is easy, as usual:

```
import tensorflow as tf
import keras
```

The simplest way to build neural networks in `keras` is by deploying different layers in a sequence. So far, we have only explored fully connected layers (also known as *dense layers*) and dropout layers, so we can begin by importing those:

```
from keras.models import Sequential
from keras.layers import Dense, Dropout
```

### 8.4.1 Case 1: How many top quarks?

We begin with the 4-top vs t-tbar problem, and use the data configuration after feature engineering (where we added the features "number of leptons," "number of jets," etc.). For reference, the optimal SVM achieved 94–95% accuracy; we want to check whether we can beat this performance with our deep learning model. The brilliance of neural networks comes from their flexibility and their ability to automatically engineer features by virtue of their layers; these qualities, albeit in lower measure, are also shared by SVMs. Therefore, we would expect our network to perform at least as well, and possibly (but not obviously!) better.

Our first step will be to think about model architecture. The input layer has 24 neurons (the number of features we have). The output layer has one neuron; the output is the probability that the object belongs to the positive class, and we can use a sigmoid as the final nonlinearity, as we know that it is suitable for predicting probabilities. In a multiclass problem, we would set up the output as $N$ neurons, where $N$ is the number of possible classes, and use softmax, which is the generalization of the sigmoid to $N$ probabilities that add up to one, as the final nonlinearity.

To establish some benchmark, let us add two hidden layers and give them both 20 neurons (the parameters of the network, e.g., the number and size of layers, should be optimized through CV). We will also reserve the possibility of adding a dropout layer after each dense layer; the dropout fraction(s) should also be optimized through CV.

Other decisions that we have to make are: which nonlinearities to use (we can start with Rectified Linear Unit, or ReLU, for hidden layers, and sigmoid for the final one), which optimizer to use (Adam), which starting learning rate to adopt (here 0.001, but again this should be decided through CV), the number of epochs (e.g., 100; we can plot the evolution of quantities of interest to check that we have enough), the batch size for the gradient descent step (here 200, but it should be explored), and the loss function. The latter is the binary *cross entropy*, or logistic loss, which is the standard choice for classification problems that output a probability $\pi_i$ that a certain example $y_i$ will belong to the positive class:

$$\mathcal{L}_{\text{CE}} = -\frac{1}{N}\sum_{i=1}^{N} y_i\, log(\pi_i) + (1-y_i)\, log(1-\pi_i). \tag{8.5}$$

Note that aside from the average and the negative sign, this loss function is the logarithmic likelihood introduced in Chapter 5 when we talked about linear models.

The cross-entropy loss function rewards not just a correct prediction but also the degree of "confidence" in a correct prediction (i.e., a high probability); the negative sign ensures that the loss has the correct behavior (i.e., that it increases for poorer predictions). For probabilities that are asymptotically close to the real labels, this loss function tends to 0, and it increases as the predictions deviate from the truth.

The following lines of code specify and build the architecture of the network:

```
model = Sequential()
model.add(Dense(20, activation='relu', input_shape=(24,)))
model.add(Dense(20, activation='relu'))
model.add(Dense(1, activation='sigmoid'))
```

As we can see, it is sufficient to specify the size of each intermediate layer and the associated activation function. We can also specify the optimizer and the loss function, which are provided through the `compile` method:

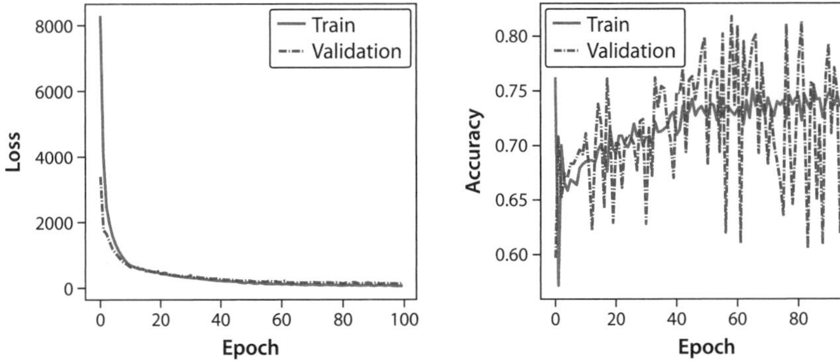


**Figure 8.8: (Left) Train and validation loss for the first model described in Section 8.4.1. (Right) Train and validation accuracy for the same model.**

```
optimizer = tf.keras.optimizers.Adam(learning_rate=0.001)
model.compile(loss='binary_crossentropy', optimizer = optimizer, metrics =
['accuracy']
```

All of these operations are described in more detail in the notebook "FirstNN_ParticlePhysicsID.ipynb." The `metric` keyword specifies which additional quantities to monitor during the training process (the network still *optimizes* the specified loss function through its Adam gradient descent). In this case, I specified accuracy, because it is more interpretable than the loss itself, which doesn't have an absolute significance; having more sanity checks is always a good idea! If you want to explore more options, the methods `keras.losses`, `keras.activations`, and `tf.keras.optimizers` contain a variety of possible choices.

Before starting the training process, we need to set up a CV strategy. For simplicity here, we will go with the simplest option, by shuffling the data and dividing them in three splits: train, validation, and test. This choice preserves the essential feature of nested CV, which is the separation of training, parameter optimization, and testing process, but is of course less than ideal, because it corresponds to one single realization, so there is no notion of *typical* behavior and no estimation of uncertainty.

It is finally time to train our first neural network. The model created above has the "`fit`" method just like any other model in `sklearn`, and we can specify the training and validation data, as well as other arguments:

```
mynet = model.fit(X_train, y_train, validation_data= (X_val, y_val),
epochs = 100, batch_size=200)
```

At the end of the training process, we can use the "`history`" attribute of the `mynet` object to check how the train loss and the validation loss have changed throughout the training process. This is shown in the left panel of Figure 8.8.

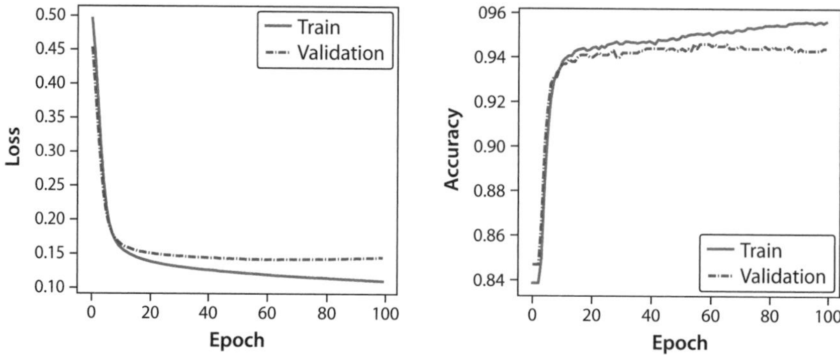


**Figure 8.9: (Left) Train and validation loss for the second (scaled) model described in Section 8.4.1. (Right) Train and validation accuracy for the same model.**

If we were to look only at this plot, we'd probably conclude that everything looks good: The train and validation losses are similar to each other, and they start up high and gradually decrease, flattening after 20 or 25 training epochs. However, a look at the accuracy curves, in the right panel of the figure, reveals a much less rosy scenario: The training accuracy and the validation accuracy are pretty low (in the 70–75% range, while SVMs could achieve an accuracy of 94–95%), and the validation accuracy is highly unstable, varying wildly from epoch to epoch. This plot is one of the reasons I always try to monitor additional indicators besides the loss.

Something is clearly going wrong with our network. It might be that we need to change something in the architecture, or train for a longer number of epochs, but before considering these options, it's good to go back to the checklist in Section 8.3.5 to see whether we have missed something. And in fact, the very first step, preprocessing, is something we haven't considered: We didn't check whether our data needed scaling. A quick look at the data confirms that different features span a wide range of several orders of magnitude, something that is harmful to models that combine different features together, like neural networks.

After applying any scaling method (here I have used `StandardScaler`, which might not even be optimal), we can use exactly the same model to obtain the loss/accuracy curves of Figure 8.9. These are much better behaved: We see some hints of a variance issue, because of the widening gap between train and validation scores, but the instability is gone, and the validation accuracy has much more reassuring values, hovering around 94%. This is a well-behaved model that we can try to further optimize with some of our usual tricks. This example was a straight stream of consciousness from my own experience setting up this exercise: I made the mistake of not checking on data before starting out, and I wasted some time on useless optimization strategies *before* remembering my foundations and doing my checks. This approach is particularly dangerous with neural networks, where one has so many knobs to turn and options to try; it's easy to get lost in the hyperparameter

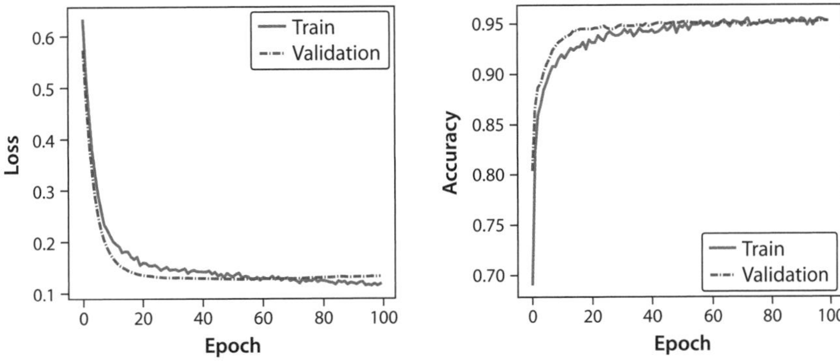


Figure 8.10: (Left) Train and validation loss for the third (scaled, regularized) model described in Section 8.4.1. (Right) Train and validation accuracy for the same model.

optimization process and waste a lot of (machine and human) time if those ideas are ill guided. In short, I recommend thinking before tinkering!

Having found our big issue doesn't mean that our work is done: We should explore different configurations of hyperparameters to see whether we can improve the validation results (and then quote the performance on the *test set* as our final estimate of the generalization error). Let us only look at one possible improvement: Inspired by the variance issue shown by the curves of Figure 8.9, we can try to apply some regularization mechanism. The simplest and often most effective one is dropout, which can be implemented by using the namesake layer in `keras`. We can add a `Dropout` layer after every `Dense` layer in the model we already have:

```
model = Sequential()
model.add(Dense(20, activation='relu', input_shape=(24,)))
model.add(Dropout(0.2))
model.add(Dense(20, activation='relu'))
model.add(Dropout(0.2))
model.add(Dense(1, activation='sigmoid'))
```

The code above assumes a dropout fraction of 20% for all layers. The effect of applying this simple regularization strategy is shown in Figure 8.10, which shows how the validation and train scores have converged, although the actual validation scores are essentially unchanged. This suggests that the regularized model is preferable, since we are obtaining similar validation scores but are gaining robustness to generalization.

The last step in the model evaluation pipeline is to check the final model's performance on the test fold we reserved at the beginning of this example. This test fold is as close as possible to new data, in the sense that objects in the test fold have not participated in the model optimization or training process; however, these objects were still part of the same learning set (more specifically, sampled from the same distribution). It is my opinion and experience that truly new data are hardly ever exactly akin to the objects in the test fold. This is an important caveat, which I mentioned at the very beginning in Chapter 1 and will repeat now: The test error is the best indicator that we have of the true generalization error (error on new data), but it still yields, in general, an optimistic estimate. Only if we are confident that the statistical properties of new data and test data are exactly equivalent are we allowed to fully "believe" the test error.

In this case, the test error I obtain is $\sim 0.17$, and the test accuracy is 93.5%. These are slightly worse than the validation loss and accuracy, which were $\sim 0.14$ and 94.5%, respectively. Note that we can't tell whether these differences are statistically significant, because we don't have a way of associating an uncertainty to these numbers. These uncertainties would be available if we had estimated the training and the validation errors across multiple folds; this is a good reminder that any optimization problem, including this one, would be better explored through nested CV. Overall, we can still say that the performance of this neural network is pretty close to what we obtained using SVMs in Chapter 4.

This concludes our minuscule foray into this problem's model optimization. In the exercise section, you will find some ideas for other aspects to improve.

### 8.4.2 Case 2: Photometric redshifts

In this example, we revisit the photometric redshift problem from Chapter 6, and we focus a bit more on hyperparameter optimization using the package `keras tuner`. You can follow along in notebook "FirstNN_PhotoZ.ipynb." We begin by loading the same final set of training data used in Section 6.4.2. This time, let's remember to scale the data from the beginning and set up a neural network with two intermediate layers, just as we did above:

```
model = Sequential()
model.add(Dense(100, activation='relu', input_shape=(6,)))
model.add(Dense(100, activation='relu'))
model.add(Dense(1, activation='linear'))
```

Note that, compared to the example in Section 8.4.1, we have changed the number of input neurons to match the number of initial features (6 in this case), we are using the MSE as our loss, and we don't have a nonlinear activation in the last layer, as is common in regression problems.

Our first iteration produces decent-looking train/validation curves, as shown in the left panel of Figure 8.11. Note that unlike in classification problems, where we could monitor other metrics such as accuracy or precision, we don't immediately have an interpretable quantity to look at. We can, however, make a simple scatter plot of true vs estimated redshift, shown in the right panel of the same figure; this is a good sanity check that our predictions are at least reasonable.

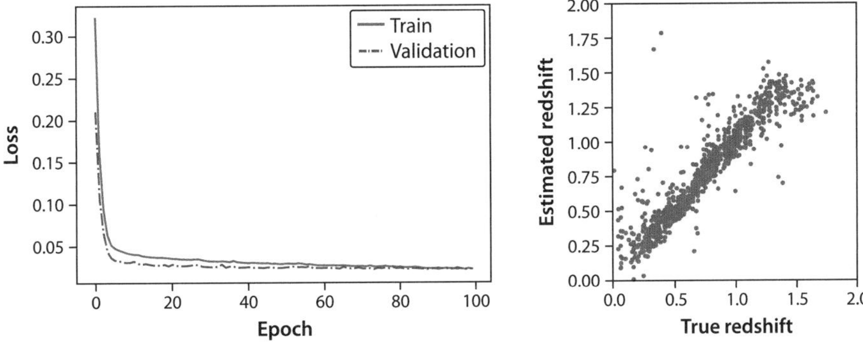


**Figure 8.11: Train/validation curves (left) and predicted vs ground truth values (right) for the photometric redshift problem, from the first neural network described in Section 8.4.2.**

If we are interested in other performance indicators beyond the loss, we can potentially define an ad hoc metric to monitor during the training process or just calculate it at the end. In our case, we are interested in the normalized median absolute deviation (NMAD) and the number of outliers, defined as those objects for which $\Delta z/(1 + z_{\text{true}})$ exceeds 15%. When we calculate these values on the test fold, we find NMAD = 0.037 and OLF = 0.047 (note that we only have a single-point estimate), which are only slightly worse than the optimal numbers obtained with bagging and boosting algorithms.

Our challenge will be to improve on those numbers using hyperparameter optimization.

We are particularly interested in exploring the effect of the size of the network, both as width and depth, as well as in understanding how the choice of different loss functions affects the derived metrics of interest (NMAD and OLF).

### 8.4.3 Hyperparameter tuning in `keras`

Several approaches can be used to tune the hyperparameters of a neural network. My recommendation is to play with them first and use the insights obtained to decide which hyperparameters are worth tuning and which ones have a relatively small effect on the overall performance.

Assuming that we have reduced our search space, perhaps the simplest way to proceed is to use one of the packages that are built for hyperparameter optimization, for example, `keras tuner`. When using one of these tools, we give up some flexibility in parameter tuning but gain ease of use. As a result, I consider them a great starting tool.

The two ingredients that we have to provide are the `keras` model, which will now include some "fixed" bits and some "variable" ones, and the tuner itself, which specifies how to explore the hyperparameter space. The model can be specified as a class or a function; the latter is the simplest option, and the one we examine here.

In this example, we will only vary the number of layers in the network, the size of each layer, and the learning rate. I mentioned previously that the Adam optimizer is quite efficient in selecting a suitable learning rate, but technically this is only valid when the number of epochs is adjusted as well.

The resulting model could look like this:

```
def build_model(hp):
    model = keras.Sequential()
    for i in range(hp.Int('num_layers', 2, 6)):
        model.add(layers.Dense(units = hp.Int('units_' + str(i),
        min_value = 100, max_value = 500, step = 100), activation = 'relu'))
    model.add(Dense(1, activation = 'linear'))
    model.compile(optimizer = tf.keras.optimizers.Adam(hp.Choice('learning_rate',
    [1e-2, 1e-3, 1e-4])), loss = `mse')
    return model
```

Note that there was no specific reason to choose a maximum of 6 layers and to vary the size of each between 100 and 500 neurons, but my empirical assessment of the level of difficulty of this problem is "not very difficult," and most regression problems can be satisfactorily approached with a modest-size network (a few hidden layers, a few hundred neurons). Of course, if our parameter search indicated that the widest or deepest available network is a strong favorite among models, we should consider broadening the hyperparameter search and including larger layer sizes or higher depths.

The second ingredient is the tuner, which contains the instructions for how to sample hyperparameters, and which function to use for evaluation purposes (typically, the validation loss). Note that unlike in some other ML algorithms, every time we train a neural network, the outcome is not entirely deterministic, because the weights' initial values are generated from a random distribution, and because of the stochastic nature of batch gradient descent. Therefore, it is advisable to evaluate each model more than once; this is the meaning of the "executions_per_trial = 3" parameter. Finally, the "max_trials" parameter specifies how many combinations will be explored through the random search; this number should be large enough to yield a high probability of identifying the most interesting combinations but significantly smaller than the total number of models described by a grid approach. Our tuner looks like this:

```
tuner = RandomSearch(build_model, objective='val_loss',
max_trials = 40, executions_per_trial = 3, project_name = 'Photoz')
```

Just as we did in Section 6.6.2, we use a random search approach rather than a grid search. This helps limit the computational expenses in training a large number of neural network models, but there are even better approaches. For example, `keras tuner` offers `Hyperband`, an optimized random search where a larger fraction of models is trained for a smaller number of epochs, but only the most promising ones survive, and Bayesian optimization, which attempts to build a probabilistic interpretation of the model scores as a function of the hyperparameters, modeling the scores as a Gaussian process in hyperparameter space (e.g., [Brochu et al., 2010; Snoek et al., 2012]). The recent review [Yu and Zhu, 2020] is a good resource for hyperparameter optimization strategies.

Once the hyperparameter space has been defined, we can deploy the tuner similarly to a `keras` model:

```
tuner.search(Xst_train, y_train, epochs = 100,
validation_data = (Xst_val, y_val), batch_size = 300, verbose = 0)
```

The tuner runs the optimization by considering different random combinations and ranking them using the loss on the validation fold. An estimate of the generalization error can be obtained by evaluating the model on the test fold. As we saw when the grid search algorithm was introduced in previous chapters, multiple configurations give similar results, and these single-point estimates can trick us into thinking that small differences matter more than they do. So I definitely recommending investing resources in a CV approach (discussed at the end of this section) before believing that a more expensive network (wider, deeper) is really crucial for your task.

In our case, the results of the tuner search give us some interesting insights:

- The difference in scores among the top few models (say the top five) is minimal.
- The initial learning rate matters; all the top-performing models have a learning rate of 0.001.
- There are similar configurations that give similar results; our top five includes two 3-layer models, two 4-layer models, and one 5-layer model. We will pick the first, because it corresponds to the smallest number of total trainable parameters.
- The size of the hidden layers also varies significantly within the top five models, indicating that several configurations are equivalent. Again, for simplicity, we will just pick the configuration of the top model (400 - 100 - 500 neurons), but a quick look at the other models suggests that the "hourglass" shape of our model is not a meaningful feature.

For this particular model, the performance of the optimized neural network yields a slightly higher test MSE loss (0.04), but better secondary indicators, NMAD and OLF, which are 0.031 and 0.029, respectively. Again, without a cross-validated approach, it is hard to know whether these improvements are statistically significant.

Our last exercise in optimization has to do with customizing loss; more precisely, considering how choosing different loss functions may affect our secondary metrics (in this case, the OLF and NMAD). I left this for last, but this is possibly my favorite argument in support of neural networks: we can choose whatever loss function we want! This is very different from what happens even in the most flexible and powerful algorithms we have seen so far, for example, SVMs or tree-based ensembles. For all of those algorithms, we would specify a metric to monitor, but at most, this information was used when *comparing* models (e.g., during hyperparameter optimization). It was never used in actually *building* the model: In SVMs, the optimal model is always built by maximizing the width of the margin (adjusted for regularization, of course), which only depends on the support vectors. In tree-based methods, the optimal model is the sequence of splits that maximizes the decrease of impurity on a local, split-by-split basis. With neural networks, in contrast, we have the option to choose which loss function we want to minimize during the training process. This makes these networks incredibly flexible!

In our case, we will now stick with the optimal architecture found above and consider three possible loss functions (out of the many ones available in `keras`): the mean square error, which was our current choice, the median absolute error, and the median absolute percentage error. Comparing the three losses directly is not feasible, of course, but we can compare our derived metrics and pick the loss that gives us the best results. Within `keras`, we could also directly define a custom loss.

In the notebook "FirstNN_PhotoZ.ipynb," we train the network on the same train/validation/test splits, using the three losses specified above. To obtain some sense of how much the derived metrics might fluctuate as a result of different network initialization and training processes, we run the same process a few times. Using the MSE and MAE loss tends to produce better (lower) indicators for both OLF and NMAD, although we can't (yet) quantify the statistical significance of this statement.

### 8.4.4 Test error evaluation and CV in `keras`

I have mentioned several times, in this chapter and elsewhere in this book, that single-point estimates of validation or test error might not be reliable, as they could be atypically high or low, and don't provide any assessment of statistical uncertainty. We are familiar with CV techniques in `sklearn`, but how can we use them in the `keras` framework?

We have several options, depending on what our goal is and how much flexibility we require. In my opinion, the simplest option is to "recycle" many of our trustworthy `sklearn` tools by using the `scikeras` "wrapper"[1] Application Programming Interface for `sklearn`, which allows us to define `keras` objects as regular `sklearn` estimators. In this case, since we are dealing with a regression model, we can import the relevant wrapper like this:

1 https://github.com/adriangb/scikeras

```
from scikeras.wrappers import KerasRegressor
```

The KerasRegressor object takes a function that creates a keras model as its first argument (build_fn) and returns a sklearn estimator:

```
estimator = KerasRegressor(build_fn = create_model, epochs = 100, batch_size = 200,
verbose = 0)
```

What we do next depends on whether we are optimizing *hyperparameters* or simply assessing *models*.

If we are just trying to estimate the test error of a single model, we can use a single-layer CV strategy. If we are doing any parameter search, though, we need to maintain a three-tiered strategy with a repeated train/validation/test structure. This ensures that there is no leakage of information about the hyperparameter optimization in the test scores and requires nested CV.

We can begin with the easiest case; let us assume that we are given a model with a certain architecture (say, three layers with 300 neurons each), and we want to estimate the test scores, including an estimate of the uncertainty due to different *k*-fold splits as well as the stochastic nature of the weight initialization and batch gradient descent process.

In this case, the function that builds the keras model does not need variable arguments and can be as simple as this:

```
def create_single_model():
    model = keras.Sequential()
    model.add(layers.Dense(300, activation = 'relu'))
    model.add(layers.Dense(300, activation = 'relu'))
    model.add(layers.Dense(300, activation = 'relu'))
    model.add(Dense(1, activation = 'linear'))
    model.compile(optimizer = tf.keras.optimizers.Adam(),loss='mse')
    return model
```

At this point, all we have left to do is to define the CV strategy (e.g., a simple *k*-fold CV with shuffling) and remember (from our example in Chapter 4) that when data need scaling, we should employ a pipeline to make sure that the scaling is derived on each train fold and applied to the train and test folds:

```
pipeline = Pipeline([('scale', StandardScaler()), ('model', estimator)])
```

We are now ready to employ functions like cross_validate to our pipeline object:

```
scores = cross_validate(pipeline, X, y, cv = cv,
scoring = 'neg_mean_squared_error', return_train_score = True)
```

In this case, the test scores (MSE) we obtain are $0.021 \pm 0.003$. We can use the function cross_val_predict, as we already did in the examples of Chapter 4, to derive several predictions of the target values and calculate similar estimates for derived metrics, such as OLF and NMAD.

The same framework can be used for the more challenging task of optimizing and scoring `keras` models.

For optimization, we need to combine the ability to vary different hyperparameters with the CV search strategies available in `sklearn`. The `GridSearchCV` is unlikely to be practical; the `RandomSearchCV` is a better choice. Note that we are "losing" some freedom in exploring the hyperparameter space with respect to `keras tuner`, but we gain CV superpowers, which may be worth the trade.

The next step is to define a new function to build a `keras` model and to make sure we can vary the arguments we are interested in optimizing. In this case, we are keeping the number of hidden layers at 3 and will vary their sizes as well as the learning rate. Some other parameters will be added directly as part of the parameter grid:

```
def create_model(lr = 0.01, size_1 = 500, size_2 = 100, size_3 = 400):
    model = keras.Sequential()
    model.add(layers.Dense(size_1, activation = 'relu'))
    model.add(layers.Dense(size_2, activation = 'relu'))
    model.add(layers.Dense(size_3, activation = 'relu'))
    model.add(Dense(1, activation = 'linear'))
    model.compile(optimizer = tf.keras.optimizers.Adam(learning_rate = lr), loss = 'mse')
return model
```

Next, we define our set of hyperparameters, which will be the input for our random search:

```
batch_size = [200, 300, 400]
lrs = [0.0001, 0.001, 0.01]
epochs = [50, 100, 200]
size_1 = [100, 300, 500]
size_2 = [100, 300, 500]
size_3 = [100, 300, 500]
```

Finally, we are ready to pick the CV strategy, to define the `keras` model that uses the `create_model` function above, and to run the random search. For this example, we are varying 7 parameters, and each one of them has 3 possible values, so we have a total of 729 combinations that we could potentially explore. In our random search, set `n_iter` to 40 to make sure we have a high enough probability to find good models, while keeping running time somewhat reasonable. One last important technicality is that as in any CV scheme, we need to create a pipeline in order to scale the data using only training set information:

```
kmodel = KerasRegressor(build_fn = create_model, verbose = 0)
pipeline = Pipeline([('scale', StandardScaler()), ('est', kmodel)])
```

We can access the parameters of the keras model through the prefix "est" (just the name we gave to the model), followed by "__"; this allows us to define the hyperparameter grid and ultimately run the CV:

```
cv = KFold(n_splits = 4, shuffle = True)

param_grid = dict(est__size_1 = size_1, est__size_2 = size_2, est__size_3 = size_3,
est__lr = lrs, est__batch_size = batch_size, est__epochs = epochs)

grid = RandomizedSearchCV(estimator = pipeline, param_distributions =
param_grid, n_iter = 40, n_jobs = -1, cv = cv, return_train_score = True)

results = grid.fit(X, y)
```

After this (fairly computationally intensive) process, we can take a look at the best-performing models in the results object. Typically, we will pick the winning model, but in many cases, there are several models that perform similarly, and we might choose to pick the simplest or most robust instead.

In this case, I find that the best performing model has 300 in each of the 3 hidden layers, a learning rate of 0.001, and is trained for 200 epochs with a batch size of 200 (reassuringly, these values are quite close to the single-point estimates obtained in Section 8.4.3), but a few other models perform comparably well.

This procedure settles the question of which model(s) perform(s) best, but it still doesn't provide a proper estimate of the generalization error, which should be computed on data that have never participated in the hyperparameter optimization or training process.

Providing a fair assessment of the test scores (and their uncertainty due to the stochastic nature of sample selection and the nondeterministic aspects of the neural network) requires a three-tiered structure, with two nested CV processes: the outer CV "peels out" the test folds, and the inner CV does the validation/parameter optimization (see also Chapter 4). A fully developed example can be found in the notebook "FirstNN_PhotoZ.ipynb."

Needless to say, this process is very computationally expensive; even in a minimal setup with an outer CV with 4 folds and an inner CV with 3 folds, and 40 iterations in the random search, we train $12 \times 40 = 480$ deep models! In this case, we find that the average loss (MSE) of the winning model is $0.022 \pm 0.006$; for the OLF and the NMAD, these values are $0.031 \pm 0.003$ and $0.027 \pm 0.002$, respectively. These are the final expected scores (i.e., the generalization error) on unseen examples.

This study concludes our dive into the training, optimization, and scoring of fully connected neural networks. As you might know from reading the literature or just following science Twitter, we have covered only a tiny fraction of possible neural network *architectures*! There are many excellent resources for practitioners who want to learn more about them, and the methodology pipeline we introduced

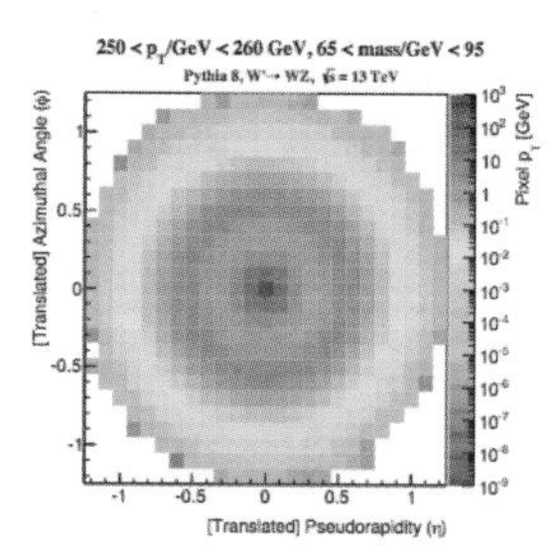

**Figure 8.12: An example of three images that can be used for facial recognition (left), particle classification (middle, from [de Oliveira et al., 2016]), and morphology classification (right, from the GalaxyZoo project.[2]). While the images are very different, for all of them, a positive outcome will depend on correctly identifying some specific shape elements in the image.**

here will be useful to train those as well. In the next section, we will browse through a small set of popular and powerful network architectures, and provide some references to learn more about them.

## 8.5 BEYOND FULLY CONNECTED NETWORKS

In my opinion, neural networks are so successful because they offer a natural way to break up very complex tasks (i.e., representing very complex functions) into smaller ones, by means of the stacked input/output structure.

In the examples that we have discussed so far, we didn't deliberately try to assign specific tasks to different layers. We can hope, of course, that optimizing the number, size, or activations of different layers would implicitly achieve this goal, but we can actually do better by employing *specialized architectures*.

These network structures either employ new types of layers or favor particular arrangements, introducing constraints between the sizes of different layers. So far, we have only considered dense layers, in which every neuron is connected to all neurons in adjacent layers, and dropout layers, which are in fact not "real" layers, but indicate that a random fraction of neurons from dense layers should be excluded at different epochs during the training process. We will take a brief look at convolutional and Recurrent Neural Networks, as well as autoencoders (and very briefly, adversarial networks).

### 8.5.1 Convolutional Neural Networks

Convolutional Neural Networks (CNNs) have been the object of well-deserved attention in the sciences, including physics and astronomy, for several years. This network architecture is particularly suitable for analyzing images. To understand why, we can think of how information is distributed in images. Figure 8.12 shows

2 https://www.zooniverse.org/projects/zookeeper/galaxy-zoo/

three images that we can associate with three different tasks: from left to right, facial recognition, particle jet classification, and galaxy morphology classification. What do they have in common? In all cases, we are trying to extract information from the image: We want to recognize the girl with the pearl earring; associate the scattering pattern to a particular type of particle; and label the galaxy as a spiral galaxy, based on its shape. These tasks are typically quite easy for people, but difficult for machines, because the information is presented in a complex way, where the ensemble is more important than the components. We understand that the brightness in each pixel is important, but no single pixel has particular importance; in other words, our initial representation with individual pixel brightness as features is not particularly helpful. Before the network can elaborate a result, we need to *synthesize* the information, extracting the elements that are relevant. Typically, we use features like edges, identifiable as sharp contrasts in luminosity between adjacent areas, or we detect features by looking for correlations at different scales. For example, when we identify "eyes" in a face, we use both information about brightness in two distinct areas and information about their orientation with respect to the rest of the image and their distance from one another. Convolutional Neural Networks have built-in layers, called (you guessed it) *convolutional layers,* that attempt to do these operations for us and return a more efficient representation of the image. The main operation in a convolutional layer is the creation of multiple *filtered* representations of the image. We can think of filters as mathematical operations that will return a non-null result if a certain feature is present. For example, an edge filter will return a positive result in the area of the images where edges are found. It is customary to use filters that are significantly smaller than the image and to "slide" them through the image, from left to right and from top to bottom, so that the filter is applied to all regions. For example, a filter might be a "tile" of 3 pixels by 3 pixels, and it will be applied several times to all the possible square areas of 9 pixels in the image, as shown in Figure 8.13. But what does it mean to "apply" the filter? At each step, we consider two matrices of weights: the filter itself, and the $3\times3$ pixel (or whatever size the filter is) area that we are considering in the image. In Figure 8.13, the subset of pixels in blue are the first ones to be multiplied by the filter: We multiply, element-wise, the two matrices, and add the results to obtain a single number. The process is then repeated, sliding the filter window to the right by one pixel, considering a new set of tiles (in the figure, the red ones), computing a new element-wise product between the two matrices, and adding the results. Once we have exhausted all possible sets, we obtain a slightly smaller matrix whose size depends on the size of the filter (in this simple configuration, the linear size will be $N-(k-1)$, where $N$ is the linear dimension of the input image, and $k$ is the linear size of the filter), which represent our first *representation* or *feature map*.

The cleverness of CNNs is that we don't have to limit our search for feature maps to one filter: We can repeat this operation multiple times, using different filters (i.e., different matrices of weights) that are engineered to capture different features

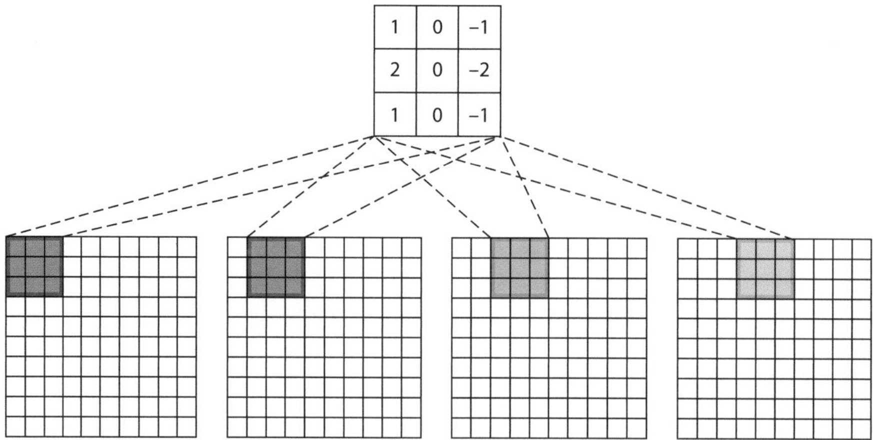


**Figure 8.13: In a CNN, different filtered maps are created by multiplying different areas of the image by small matrices ("filters") that act as feature extractors, in a sliding fashion. This is the convolution process that gives the network its name. The $3 \times 3$ filter shown here as an example is the horizontal Sobel filter, which is designed to detect horizontal gradients in an image, but the optimal filters are part of the network parameters and are obtained through the training process.**

(edges, correlations, etc.). After a forward pass through a convolutional layer, our network will have changed not just shape, but also dimensions: our original two-dimensional representation will be replaced by a three-dimensional set of images (the different feature maps), which is to say, the new representation also has a *depth* corresponding to the number of filters we have applied.

For example, in the simple example shown in Figure 8.13, where the input size is $10 \times 10$ and filters are $3 \times 3$, the final dimensions will be $8 \times 8 \times D$, where $D$ is the number of filters.

The convolutional layer, unsurprisingly, is the core element of CNNs. It comes with different variants: for example, it is customary to add "padding" to the sides, in order to avoid sharp features at the edge of images caused by an uneven number of contributing pixels in the outside areas, or to adopt *strides*, in which the filter multiplications are spaced apart by more than one pixel. Just as in any normal fully connected layer, we choose a nonlinear activation function that completes the forward pass through this layer.

It is also common to have *pooling layers*, which are meant to be a dimensionality reduction (DR) step. In pooling layers, we aggregate the information coming from (small) square blocks of pixels, for example $2 \times 2$, by taking either the average or the maximum value of the four pixels. It is, *de facto*, a rudimentary two-dimensional interpolation, which is useful for reducing the size of the network but is not favored by all practitioners (see, e.g., the discussion here[3]).

3 https://cs231n.github.io/convolutional-networks/

What happens when we stack multiple convolutional layers? Because of the pooling operation, at each step, the filters connect larger regions of the original image (or other two-dimensional input). Thus as we move forward through different convolutional and pooling layers, we create features that represent different spatial correlations over different, increasing scales. This is a unique and helpful feature of CNNs: *They have an intrinsic feature-generation process where higher-level features appear as we move through the layers of the network.*

A natural question at this point is: How should we choose the filters that describe the relevant features of the images? Well, of course, we don't: We build them through the training process! While the shape and number of filters are chosen as part of the architecture of the network, the filters themselves (the numbers that appear in the little matrices) are parameters of the network, and just like any weights and biases, they are optimized through backpropagation. Thus we only make the assumption that optimal features at different scales *exist*, but we don't predetermine what they are. After the training process, we can even visualize the filters to learn what kind of features were generated, although their interpretability becomes murkier as we move forward along the network and away from the input.

Finally, it is important to note that in most cases, the parameters that determine the filters are shared among all locations (i.e., as we move along the image capturing the convolutions, the numbers that make up the filters stay the same). This is equivalent to saying that the features we are trying to make emerge, such as edges or shapes, will be similar across the entire image, which seems reasonable under the assumption of *translational invariance* of images. This assumption greatly reduces the number of trainable parameters, although it might be limiting or not appropriate in cases where different areas of input images present very different characteristics.

To summarize, there are two tasks that CNNs can do really well: (1) create a hierarchical representation of relevant features in images, effectively generating a new representation space, and (2) incorporate information at small and large scales and use the information to reduce the number of weights in the network, increasing the efficiency of the training process. The guide in [Dumoulin and Visin, 2016] contains a clear summary of the relevant mathematics, and the corresponding github repository[4] has some beautiful animations.

It is easy to build and train CNNs in `keras` by using specialized layers for convolution and pooling. The most common ones are the `Conv2D` layer, which implements the two-dimensional convolution described in the text and takes as arguments the size of the filters, stride, and padding, and the `MaxPooling2D` layer or `AveragePooling2D` layer, which implement the pooling, with similar usage. More detail can be found in the `keras` docs.[5]

---

4 https://github.com/vdumoulin/conv_arithmetic
5 https://keras.io/api/layers/

One of the typical challenges of working with image-type inputs is defining a notion of similarity, as already noted in Chapter 7, when discussing how clustering algorithms may need to be explicitly guided through a customized distance metric to recognize similar inputs. At this point, you have probably realized that classification is basically clustering with labels, so it's not surprising that we encounter this problem again. For example, to the human eye, it's pretty clear that rotating an image shouldn't affect the outcome of the classification; we have no trouble recognizing the same image seen at a different angle or from a different orientation. Clustering algorithms that are applied to image-type inputs often need to define a metric of similarity that is rotationally invariant, and this is a nontrivial task. In classification problems, however, we are afforded an easier problem: We don't necessarily need to teach our model that rotated images are similar, but only that they need to be assigned to the same class. Therefore, we just need to be sure that the rotated images are seen and recognized. This can be achieved through a technique known as *data augmentation*. It consists of enriching the learning set beforehand by adding different transformations of input images, in order to make the model more resilient. Typical transformations that are used include flipping, rotation, translation, slight distortion, and adding noise (blurring). `keras` and `TensorFlow` both have built-in methods to perform data augmentation; for example, see this tutorial.[6]

### 8.5.2 Recurrent Neural Networks

Another specialized architecture of interest is that of Recurrent Neural Networks, or RNNs [Rumelhart et al., 1985]. Their peculiarity is that the output does not depend exclusively on the input but also on the previous values of output (i.e., the current or previous *state* of the network). RNNs are typically used when the concept of *sequence* is involved, as part of the input, the output, or both.

For example, many natural language processing applications use RNNs: Understanding or translating language is best done by considering groups of words rather than single words. Consider the task of understanding the meaning of the word "ball": It is different in the sequence "I played with a ball" and in the sequence "I went to a ball." Humans are very used to understanding homophones and homonyms *from context*: Providing a sequence instead of a single input corresponds to providing context to the neural network.

In physics and astronomy, RNNs are often connected to sequences in time rather than speech or *time series*. For example, we might decide whether the status of a detector is problematic, or whether an unusual transient phenomenon (like a planet passing in front of a star, or a supernova exploding) is being detected. In many cases, readings from detectors would point to a specific event only when the *series* is anomalous, as opposed to one isolated readout value. Similarly, we could

---

6 https://www.tensorflow.org/tutorials/images/data_augmentation

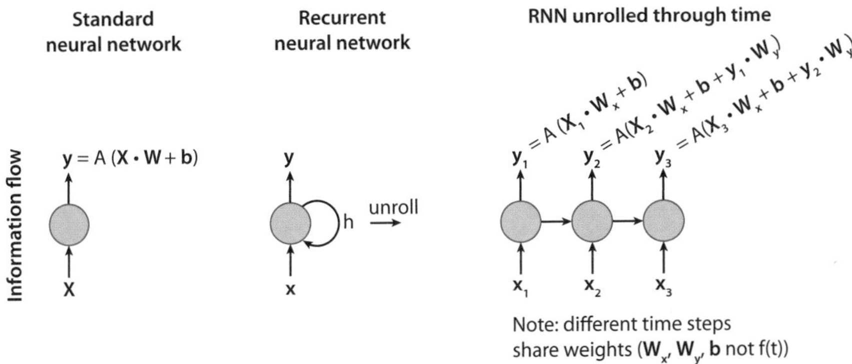


**Figure 8.14: (Left) In regular neural networks, the flow of information happens only along the input-output direction. (Middle/Right) In RNNs, information travels along two directions: The horizontal axis describes the "sequencing" dimension, for example time, and the vertical axis describes the input-output aspect, here shown for a single layer.**

imagine using our time series input to *predict* future values: This is known as *time series forecasting*. The second edition of [Ivezić et al., 2014] provides a set of nice examples.

The way we capture this information in an RNN is by adding a second dimension to the information flow: not just from input to output, but also across time. Note that (somewhat confusingly) the word "time" is generally used to denote the dimension that is associated with the sequence, but this does not imply that we are analyzing a time series; words in a sentence would still be considered to be aligned along the "time" direction.

As already mentioned, the output for each time step subsequent to the first depends not just on the input at the same time step, but also on the output (status) at the previous step. This is easy to represent in a two-dimensional diagram. Figure 8.14 shows how information flows toward the intermediate layer (here a single neuron) from the left (along the time direction) and from the bottom (along the input-output direction). Incidentally, this diagram is rotated 90° with respect to our usual representation of neural networks where the input-output flow is shown left-to-right. This orientation seems to be quite standard in the RNN literature, which is why we adopt this convention here.

We can actually write down the output of each time step, at least for the simplest neural network where the information about the current status of the network (sometimes denoted with **h**, for *hidden*) coincides with the output of the previous step.

In a regular neural network, the input **X** of a given layer generates an output **y**, given an activation function A:

$$\mathbf{y} = A(\mathbf{X} \cdot \mathbf{W} + \mathbf{b}). \tag{8.6}$$

In a recurrent neural network, *unrolled through time* (i.e., where we make explicit the dependence on time), the output will depend on a second weight matrix:

$$\mathbf{y}(t) = A(\mathbf{X}_t \cdot \mathbf{W_x} + \mathbf{b} + \mathbf{y}_{t-1} \cdot \mathbf{W_y}). \tag{8.7}$$

Note that the weight matrices are shared by all time steps i.e., they are independent of time. We have one matrix $\mathbf{W}_x$, one matrix $\mathbf{W}_y$, and one set of biases $\mathbf{b}$ for every layer of the network.

While Eq. 8.7 only makes explicit the time dependence on the output of the *previous* step, because each output depends on the previous step's output, formally the RNN *carries memory of all previous steps*. This is referred to as the hidden state of the network. However, the impact of the output of each time step gets diluted as the network is unrolled along the time direction, so that the RNN's memory effectively *fades away*. This might work in cases when information is encoded in short sequences, but it might be a problem when we want the network to remember and be sensitive to outputs from several steps earlier. Together with the exploding and vanishing gradient problem [Pascanu et al., 2013], the issue of fading memory is one of the main challenges of training RNNs.

A successful breakthrough was obtained by replacing the simplest "memory cell" of RNNs, which is the block that encodes the information from the current status, with a more sophisticated structure. In Eq. 8.7, the hidden state is simply the output of the previous step, and we effectively treat it as an additional set of neurons in our layer. Instead, solutions like Long Short-Term Memory (LSTM) cells [Hochreiter and Schmidhuber, 1997], or Gated Recurrent Units [Cho et al., 2014], help refine the way information is passed along the time direction, teaching the network to retain important memories and forget irrelevant ones.

RNNs can be successfully used for problems where a sequence is involved either as input or as output, or both. Language interpretation or translation is a typical example in which both input and output are sequences (groups of words in a given order). Detecting transient objects or detector malfunctions are examples of an input sequence (e.g., read-out values) producing a single output (positive detection or status). Finally, it is also possible to connect single-instance inputs to a sequence-type output; image captioning is an example of this type of application. These three categories of problems are known as *seq2seq* (sequence to sequence), *seq2vec* (sequence to vector), and *vec2seq* (vector to sequence), respectively, and they are illustrated in Figure 8.15. The word "vector" is used as a reminder that the input/output may not have a "temporal" (sequence) component, but that doesn't mean that it is a scalar quantity.

In `keras`, there are specialized layers that make it easy to implement RNNs: `SimpleRNN`, which implements a fully connected neural network where the output of the previous time step is fed to the next one; `LSTM`, which implements the Long Short-Term Memory cell; and `GRU`, which implements the Gated Recurrent Unit

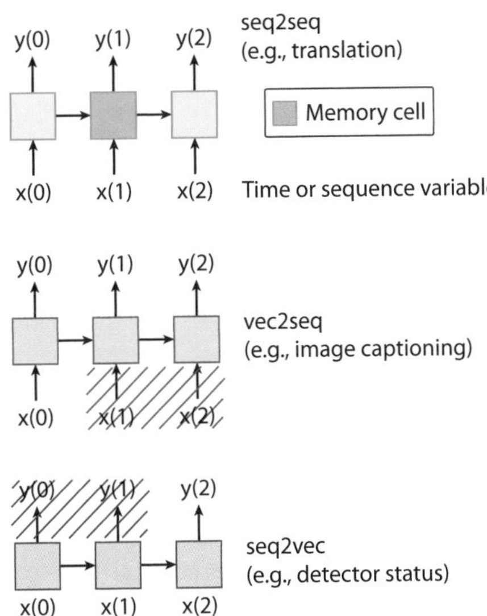


**Figure 8.15: The three diagrams correspond to (Top) *seq2seq*, (Middle) *vec2seq*, and (Bottom) *seq2vec* RNNs.**

memory cell. This tutorial[7] is a good beginner resource, while I like [Sherstinsky, 2020] for a more complete introduction to RNNs.

### 8.5.3 Autoencoders

Autoencoders are another specialized network architecture that can be used for DR purposes, but they have also recently become popular as a generative model. The network structure consists of two mirroring pieces: the encoder, which attempts to learn a latent representation of the data, and the decoder, which "undoes" the encoding operation and attempts to return a copy of the input.

The success of an autoencoder network is measured by some notion of similarity between the input and the output. The goal of the encoder is to return a perfect copy of the input. It is easy to understand that if the dimensionality of the encoder output (the "middle" layer) is the same as or higher than the input, the encoder will just learn the identity function. To encourage the network to learn a more useful latent representation, one can force the dimensionality of the encoder output layer to be lower than that of the input layer, obtaining a so-called *undercomplete* autoencoder. This scheme is shown in Figure 8.16. Another possibility is to work with high-dimensional encoding layers but enforce some type of regularization, for example, requiring sparsity in the coefficients (*sparse autoencoders*) or robustness to noise (*denoising autoencoders*). Empirically, we know that overparameterized, heavily regularized networks tend to work quite well. Regularization is a crucial step for autoencoders, because it is easy for them to overfit: A sufficiently complex

7 https://www.tensorflow.org/guide/keras/rnn

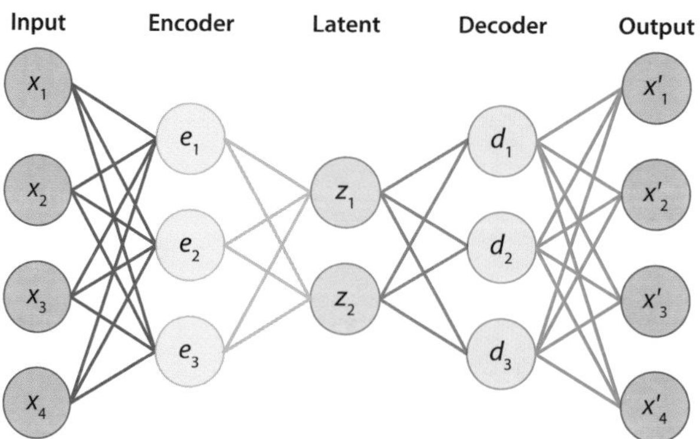


**Figure 8.16: Basic scheme of autoencoders. The encoder part (left side) tries to learn a (usually lower-dimensional) representation of the data, while the decoder part (right side) tries to revert the encoding operation and return a copy of the input. The autoencoder is successful if the output vector $\mathbf{x}'$ is close to the input vector $\mathbf{x}$. Figure reproduced with permission from [Portillo et al., 2020].**

network will learn to reproduce the input data perfectly, but it will not necessarily develop a meaningful low-dimensional latent representation. From this point of view, the process of optimizing autoencoders resembles supervised learning more than it does unsupervised learning, even if no training set or labels are required. We already encountered this concept of blending between supervised and unsupervised learning in Chapter 7, when we spoke about evaluating the performance of DR techniques. As mentioned in [Géron, 2019], we can think of autoencoders as a self-supervised learning technique. Chapter 14 of [Goodfellow et al., 2016] is a great resource to learn more about autoencoders.

A linear autoencoder with an MSE loss is equivalent to Principal Component Analysis (PCA), and will induce the same representation. Because, of course, activation functions of neural networks are nonlinear, by experimenting with different loss functions, we can hope to obtain more general low-dimensional representations.

By changing the architecture of the network, we can explore the flexibility of autoencoders. When they were first deployed, the encoder and decoder used to be symmetric in structure and shared weights. This limited the number of weights that needed to be trained. Nowadays, the symmetry is not commonly enforced, and regularization or sparsity are used to limit the freedom of the encoder and decoder. It is common to embed specialized architectures in the network structure; for example, *convolutional autoencoders* can be used successfully to learn to reconstruct image-type inputs.

As already mentioned, autoencoders can be thought of as a DR technique, if we only consider the "encoder" part of the network and use it to project inputs onto the lower-dimensional latent space. Another application that is very powerful in physics is *anomaly detection*. Because autoencoders learn to represent previously seen examples really well, they will fail on inputs that are very different (outliers,

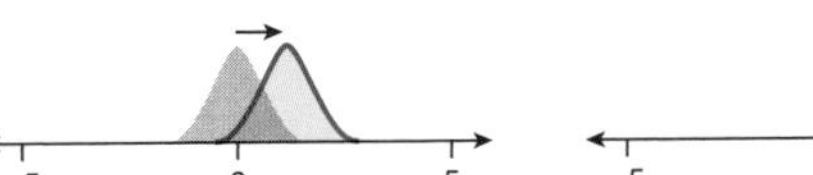
Penalizing reconstruction loss encourages the distribution to describe the input
−5
0
5
Our distribution deviates from the prior to describe some characteristic of the data
Without regularization, our network can "cheat" by learning narrow distributions
−5
0
5
With a small enough variance, this distribution is effectively only representing a single value

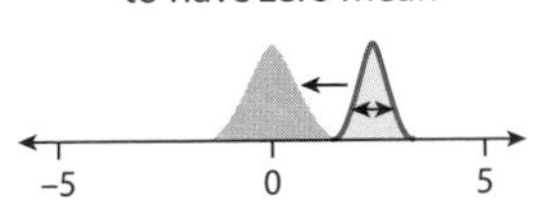
Penalizing KL divergence acts as a regularizing force
Attract distribution to have zero mean
−5
0
5
Ensure sufficient variance to yield a smooth latent space

**Figure 8.17: The role of the two components of the loss used in variational autoencoders. Reproduced with permission from jeremyjordan.me.**

or anomalies). The reconstruction error of a trained autoencoder can be used as a "flag" for anomaly detection. This technique is widely used, for example, in identifying signal (very rare) vs background (very common) events in collider physics; see, for example, [Farina et al., 2020] and many other examples listed in [Feickert and Nachman, 2021].

It is worth mentioning another specific category of autoencoders, *variational autoencoders* (VAEs) [Kingma and Welling, 2013], which in recent years has become really popular as a generative model. Recall that such models, which were introduced in Chapter 7 when discussing Gaussian mixture models (GMMs), are able to generate new input samples that are, ideally, statistically indistinguishable from the training data. In variational autoencoders, the latent space is made up of distributions rather than single neurons. We need to assume a prior for the distribution of variables in the latent space; a Gaussian is a common choice. Each neuron in latent space is replaced by two neurons, one containing the mean of the distribution and one containing the (log of the) variance. Before the decoding step, the network samples a value from the latent space and then proceeds with the decoding process, trying to reconstruct the input just as before. The loss function is made up of two parts. The first is some measure of similarity between the input and the output (a.k.a. the reconstruction loss of "regular" autoencoders), which encourages the output corresponding to each encoding to be "close" to the initial input. The second part is the Kullback-Leibler divergence between the prior distribution of the variables in latent space and the learned representation of the latent space, based on our training domain. This choice ensures that the distribution of encodings associated with a given input follows the variance observed in the training data. This second bit gives us the generative model, in the form of a probability distribution in latent space that can be sampled to generate data that are similar to the training domain. The role of the two components of the loss is shown in Figure 8.17. You can find a nice tutorial about VAEs in [Doersch, 2016].

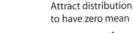

Finally, we can briefly discuss another popular generative model, Generative Adversarial Networks (GANs) [Goodfellow et al., 2014]. They are composed of two distinct and competing networks: the *generator*, which tries to learn to generate realistic inputs; and the *discriminator*, which tries to distinguish between real inputs (data) and fake inputs (generated by the generator). Like VAEs, GANs are a "young" algorithm, born in 2014, but despite their young age, their remarkable ability to generate realistic inputs, especially in image form, has led to explosive popularity, such that many variations exist (see, e.g., "The Zoo of GANs," a repository[8] dedicated to all named GANs).

### 8.5.4 Uncertainty estimation in deep networks

Despite being beyond the scope of this introductory book, it is worth mentioning that one additional advantage offered by neural networks is that the topic of (epistemic) uncertainty quantification has been further developed than for other algorithms; see for example [Gawlikowski et al., 2021]. A popular option is Bayesian neural networks (e.g., [Mullachery et al., 2018; Wang and Yeung, 2020]), where we assume some priors on the distribution of weights and do probabilistic Bayesian inference on the weight *distribution*, which can be turned into a probabilistic uncertainty on the predictions. Because of the intractability of exact Bayesian inference for neural networks, approximate Bayesian inference approaches, such as Monte Carlo dropout [Gal and Ghahramani, 2016], have been shown to be successful. In Monte Carlo dropout, dropout is used not just at training time as a regularization technique but also during the prediction phase before every weight layer, with the objective of generating a distribution of predictions that can be used to derive summary statistics. Finally, another possibility is ensembling of deep neural networks [Lakshminarayanan et al., 2016]. This approach, which builds on the idea introduced in Chapter 6 that ensemble predictions might be more robust than single-model predictions, consists of training each neural network with a modified loss function aimed at estimating predictive uncertainty and ensembling the results to obtain a distribution. This github repository[9] contains an updated list of references for uncertainty quantification for deep learning methods.

## 8.6 LESSONS LEARNED

This was another long chapter, where we discussed one of the most powerful supervised learning algorithms, neural networks. They have become immensely popular in physics and astronomy, because of their ability to generate meaningful representations of high-dimensional spaces, their role in anomaly detection, and their

---

8 https://github.com/hindupuravinash/the-gan-zoo

9 https://github.com/ahmedmalaa/deep-learning-uncertainty

usefulness as simulations toolboxes. The flexibility afforded by neural networks is impressive: They can represent any function (Universal Approximation Theorem), the loss can be fully customized even at the training stage (unlike in other algorithms where the training loss is "baked" into the algorithm with only small modifications possible, e.g., tree-based methods or SVMs), and the network architecture can be customized to fit different data types, returning a high-efficiency model. However, neural networks are expensive to train, and models tend to have low interpretability; it is hard to fully understand a network's decision process. In more detail:

- The building blocks of neural networks are the dense layers, which are the superposition of a linear model and a nonlinear activation function. We learned how to build models in `keras`.
- A neural network model is trained through an iterative two-step process, composed by a forward-pass (predictions) and backpropagation. In the latter step, the chain rule is used to "distribute" gradients along the layers of network, while adaptive versions of gradient descent (e.g., Adam) are used to move toward the minimum of the loss function.
- We highlighted the importance of scaling data and discussed diagnostic tools and reparation strategies, including smart initialization of weights, early stopping, dropout, and L1/L2 (Lasso/Ridge) regularization.
- A closely connected topic is *hyperparameter* optimization. This is no small feat for a neural network, where the number of hyperparameters is potentially infinite; from the shape and size of the network to the activation function, layer structure, dropout fractions, number of epochs used in training, tolerance, or type of optimizer. It is useful to first get a sense of which parameters play a major role in order to reduce the size of hyperparameter space that we need to explore by using a random search or other method. While nested cross validation should be used to properly estimate the generalization error of the network and its uncertainty, a single train/validation/test partition is often sufficient for the purpose of optimization.
- As a general rule, many practitioners now advise practitioners to achieve robustness (low variance) by starting with a large network, and applying regularization (e.g., [Géron, 2019; Hogg and Villar, 2021]).
- We reconsidered one classification and one regression problem from the previous chapters and showed how one can build successful neural network models, using the simplest example of a fully connected network.
- Finally, we discussed some examples of specialized networks: convolutional networks, tailored to handle image data; recurrent networks, which can process sequences both as inputs and outputs; autoencoders and variational autoencoders, which create a latent representation of the data and can be used as a generative model, respectively.

## 8.7 REVIEW AND DISCUSSION QUESTIONS

Note: Questions and exercises marked by ** are more complex, open-ended, or time consuming. Those marked by * have more than one correct answer.

**Exercise 1.** * Which of these models can solve the XOR problem?

A) Linear regression
B) Single-layer Perceptron
C) Multi-layer Perceptron
D) Kernel SVM

**Exercise 2.** * What are some good aspects of neural networks?

A) They are fast to train.
B) They can approximate any function.
C) Their decisions are highly interpretable.
D) They can be customized and are very flexible.

**Exercise 3.** * Which of these techniques is used to solve high-variance issues in neural networks?

A) Nonrandom weight initialization
B) Early stopping
C) Adaptive learning rate
D) Dropout

**Exercise 4.** The Adam algorithm is used to...

A) Compute the gradients
B) Optimize the steps in gradient descent
C) Optimize the initial choice of weights
D) Estimate the scores in a model-independent fashion

**Exercise 5.** Match the input data structure and goal with the network architecture that may be most helpful. Note that this is a simplification! This question can also be used for general discussion.

| | |
|---|---|
| A) Recurrent Neural Network | 1) Image classification |
| B) Fully Connected Neural Network | 2) Time series input |
| C) Convolutional Neural Network | 3) Regression on tabular data |
| D) Autoencoder | 4) Anomaly detection |

**Exercise 6.** In a CNN with 40 filters that are $3 \times 3$, no padding, and an input size of $20 \times 20$, what would be the shape of the network after the first convolutional layer?

A) $20 \times 20 \times 40$
B) $400 \times 40$
C) $18 \times 18 \times 40$
D) $20 \times 20 \times 9$

**Exercise 7.** What is different in the training process of autoencoders, compared to other specialized architectures like CNNs or RNNs?

A) The shape of the input
B) The algorithm used for backpropagation
C) The number of layers
D) The loss function

## 8.8 PROGRAMMING EXERCISES

**Exercise 1.** Consider one of the problems discussed in this chapter, and try a different architecture of your choice (e.g., a convolutional network) on the same problem. Compare its performance as well as the training times to the fully connected network presented in the relevant notebook. What model architecture would you recommend?

**Exercise 2.** ** Consider one of the problems discussed in this chapter, and train a few fully connected networks exploring the trade-off between width and depth of the layers. Are training times only dictated by the total number of parameters (weights)? Discuss your findings about the trade-off.

**Exercise 3.** ** This Kaggle competition[10] contains more than 60,000 images of galaxies, together with a set of labels that are associated with their morphologies. Work with the first label, which corresponds to three simple classes: smooth, showing structure like disks, and star/artifact. (1) Download the data set and select a random set of 5,000 galaxies with a fixed random seed. (2) Build a simple CNN to classify the main morphology class of the galaxies. (3) Implement some data augmentation techniques and report on any improvement, or lack thereof.

10 https://www.kaggle.com/c/galaxy-zoo-the-galaxy-challenge/data

CHAPTER NINE

# 9 Summary and Additional Resources

*If this book taught you something (in jest),*
*I would hope it's to train and then test,*
*To cross-validate more,*
*To evaluate, not just score,*
*And to know data-driven is best.*

If you managed to get to this point, congratulations on your tenacity and motivation! I thought that it might be useful to have a short summary of what we have discussed and to list some additional resources to help with the next steps. My hope is that this book (and the additional materials cited in it) have provided a foundation that will allow you to use machine learning (ML) tools in your data-driven adventures. Perhaps the most useful outcome would be some understanding of how to associate a problem with a method, in the spirit of this popular diagram[1] by Andreas Müller. I will do my best to provide a similar pipeline, perhaps with a few more details and caveats, in this chapter.

## 9.1 HAVE PROBLEM, HAVE DATA: WHAT NEXT?

Let's say that we have an open question, and some data that we hope to use to answer the question. What shall we do next? Clearly, there is no unique right answer, but we can at least try to come up with a reasonable sequence of operations that might help us find some focus.

### 9.1.1 Step One: Data exploration and manipulation

The first step in our pipeline is a humble but often crucial endeavor: exploring the data to gather some insights and "prepare" them for use. I almost always start with some sort of visualization (a simple histogram plot of each of the features is helpful)

1 https://peekaboo-vision.blogspot.com/2013/01/machine-learning-cheat-sheet-for-scikit.html

and proceed from there. Table 9.1 lists some questions that I have found useful to ask, together with a few suggestions for how to answer them and some proposed actions.

### 9.1.2 Step Two: Problem setup

Now that we understand the data better, it is time to decide how to set up our problem (note: we might need to do this more than once, if things don't work out the first time!). The first question, whose importance I can't stress enough, is simply: *Is machine learning the right way to go about solving this problem?* Despite our obvious affection for ML tools and our many attempts to open the black box throughout this book, there are many contexts in which classic inference or statistics are a better choice. Examples include: (1) Models for which the physics is very well understood, and we can be confident that we chose well-motivated parameters and sensible priors, so a probabilistic inference approach would be better; (2) Small data sets (in particular, data sets with very few features), where creating and testing simple models may be easy; and (3) Problems where physical interpretability is paramount. If we are convinced that ML methods are the way to go, the next question is: *Supervised or unsupervised?* There is often overlap between the two, and multiple possible solutions to the same problem exist, so the answer may not be clear cut. Let's consider a few possible project types:

- *Dimensionality reduction.* This is one of my favorite examples, because it's the "textbook" example of unsupervised learning, yet it can be cast as either a supervised or unsupervised method. Some unsupervised DR techniques, for example PCA or clustering, are agnostic to the problem we want to solve and tend to be robust (if we later want to use the same data to solve a different problem, we don't need to change anything). However, moving to a different feature space means that we lose the interpretability of our features. Conversely, "supervised" DR techniques, for example Lasso regularization or feature selection based on feature importance, score higher in interpretability and accuracy (because they are problem specific, we need to build a supervised model before we can use them), but are less generalizable. What will work for you, you alone can tell.
- *Data visualization/idea testing.* Suppose we have some data (e.g., observations of particle trajectories, or galaxy spectra, or weather patterns) and some ideas to test, or maybe we are just hunting for trends. For the sake of simplicity, let's say that we want to group together observations that are similar and then figure out what makes them similar. In this case, we could use an unsupervised clustering method first and then ask whether the groups formed have something in common. We could use one of the clustering methods from Chapter 7, like $k$-means, but the resulting clusters live in the same-dimensional space of input features, and they may be hard to explore. An often better approach is to use

Table 9.1: **A set of possible questions to ask in our Step One: data exploration and manipulation.**

| Questions | How to answer | Possible actions |
|---|---|---|
| Are all the features of numerical type? If not, what transformation should we use? | If we read the data in a data frame called `df`, we can simply use `df.dtypes` to obtain a list of data types for each column. | For Boolean variables, the easiest fix is to use `LabelEncoder` to turn them into 0/1. For categorical variables with more than one possible value, we can also consider one-hot encoding with `OneHotEncoder`, which avoids creating an unphysical distance metric among categories. |
| Are all the features in a similar numerical range? | For a data frame `df`, `df.describe` provides summary statistics for all numerical columns, including mean, median, and a few significant percentiles. | You might decide to normalize/standardize the features now, but I recommend waiting until you have decided which method you will use. |
| Are many of the features highly correlated with one another? | You can use the `corrcoef` from `numpy` on the feature matrix, or `pandas`' `df.corr` on a data frame `df`, to visualize the correlation coefficients among each pair of features. | If many features are highly correlated, it may be a sign that you will be able to reduce the feature set without loss of information, or that you might need to apply some regularization later on. |
| Are there many outliers, and if so, what should be done with them? | If the data set is not too large, I like to plot the distribution of each feature (non-pro trick: if you don't set the range of the histogram and it looks much larger than it should, it means that there are some outliers at the very end). As an alternative, you can compare the summary statistics in `df.describe`; a diverging mean and median point to a skewed distribution, and the min and max of each feature may also carry information. A quick way of flagging outliers is to find objects whose feature values lie more than a few standard (or median) deviations from the mean (or median), as discussed in Section 3.1. | What to do really depends on what the presumed *source* is for outliers. Are they interesting (but weird) objects? In this case, you might want to keep them. Are they most likely the result of a malfunction in the data collection process? Then usually they should be discarded. |
| Are there missing data? How many? Are they focused on a few features or uniformly distributed? | In `pandas`, it's easy to find missing data using `df.isna()` on each column of a data frame `df`. | Missing data usually require an imputing strategy. The simplest ones, like replacing a missing value with a constant value or the mean of the feature, are usually fine if the fraction of missing data is very small, but you will need something more sophisticated (see, e.g., Section 4.3.1), if the fraction of missing data is significant. |
| If you have a target (output) property, what does its distribution look like? | You can just plot a histogram of the output variable. | You probably don't want to take action at this time, but keep this information in mind for Step Three (Section 9.1.3). For a continuous variable, a very skewed distribution might be helped by a transformation (e.g., `np.log`); for a discrete variable or class, a very imbalanced distribution might be relevant when choosing a metric, selecting hyperparameters, or resampling. |
| Are the data too big to use? | This really depends on your machine and your time constraints; you might know it by this stage, or find out later once you set up a method and find that it would take too long to run your analysis pipeline. | Dimensionality reduction has many facets; we discuss it more in detail in Step Two (see Section 9.1.2). |

something like a Self-Organizing Map or a t-distributed Stochastic Neighbor Embedding map. These maps group together similar objects but also provide a two-dimensional map where similarity in the higher-dimensional feature space is mirrored by adjacency in the two-dimensional space. We could then "color" the elements of the map according to different properties. A visually detectable gradient would signal that the property we are mapping through color is highly correlated with similarity, and would suggest further exploring this idea (e.g., by creating a predictive model). This example is one of many where unsupervised learning and supervised learning can blend, which makes for the most fun applications.

- *Predictions.* This is an easy case: We know that we'll be reaching for a supervised learning model. However, plenty of choices still remain to be made: Will it be a classification or regression problem? When the output is a continuous variable, the choice of regression is obvious, but we have seen many cases in which classification problems can be recast as regression problems by predicting the probability that an object belongs to a class. If possible, I recommend the latter approach as it retains the degree of confidence we have in our classification. The next obvious problem is: What algorithm should we use? This is a very tricky problem, so I'll give you my cheat answer: Start with a Random Forest or GBM. They are fast, interpretable, and often flexible enough; if not, upgrade to SVMs; if not, go neural. Table 9.2 summarizes the pros and cons of different algorithms that you can perhaps use as a starting point. And of course I should mention here the very tempting `lazypredict`[2] package, which will build a whole set of models for you and return an algorithm recommendation. I have never used it myself, but it sure sounds worth a try!
- *Data augmentation or generation.* The ability to generate data that are similar (from a statistical perspective) to those you already have is very valuable: You might use it to effectively increase your training set size and improve your model, or to create a probabilistic description of data you have, which affords you a whole new set of statistical applications. For those goals, generative models, which turn your data into a distribution from which you can draw new samples, are usually the way to go. You can start with the simplest ones like GMMs and explore more complex ones like variational autoencoders.

### 9.1.3 Step Three: Evaluation

After data exploration and problem setup, a good next step is to decide how to evaluate the ML model. If dealing with supervised learning, we will have chosen between classification and regression, and we have a choice of "standard" or custom metrics. A custom metric can be used, for example, to give a different weight to errors made

2 https://github.com/shankarpandala/lazypredict

Table 9.2: **This very qualitative table shows various properties of algorithms that we have discussed in this book. kNN = *k* Nearest Neighbors; DT = decision trees; SVM = Support Vector Machines; LM = linear models; GLM = generalized linear models; RF = Random Forests; GBM = gradient boosting machines; ANN = Artificial Neural Networks; CNN = Convolutional Neural Networks; RNN = Recurrent Neural Networks; sw = somewhat. Columns refer to interpretability, speed, ability to internally engineer features/generate a latent representation, how much flexibility in behavior can be obtained by changing parameters, to what extent the loss that is optimized in building the model can be changed by the user (in the `sklearn/keras` framework), whether the algorithm builds a local model and thus is potentially able to extrapolate outside the training domain, and preferred data type (this may read better backward, e.g., CNNs can certainly handle tabular data, but they are especially suited to process image data).**

| Algorithm | Inter-pretability | Speed | Internal Feat Eng | Flexible Params | Flexible Loss | Extrap capable | Data type |
|---|---|---|---|---|---|---|---|
| kNN | high | high | no | low | sw | no | tab |
| DT | high | high | no | med | no | no | tab |
| SVM | low | low | yes | high | sw | sw | tab/image |
| LM | high | high | no | low | sw | sw | tab |
| GLM | med | high | no | low | sw | sw | tab |
| RF/GBM | med | med | no | med | no | no | tab |
| ANN | low | low | yes | high | yes | sw | image |
| CNN | low | low | yes | high | yes | sw | image |
| RNN | low | low | yes | high | yes | sw | series |

in different areas of feature space or to define percentage-type errors that don't seem to get much love in `sklearn`.

Remember that in the `sklearn` framework, only a few algorithms, such as linear models or neural networks, allow you to fully customize the metric that is optimized in *building* the model (e.g., in choosing the coefficients of the linear model or the weights of the network). Algorithms that let you do this will usually have a "loss" attribute (e.g., `sklearn.linear_model.SGDRegressor`).

In other cases, such as tree-based methods, the loss function is hard-coded into the algorithm itself (e.g., splits are always chosen on the basis of a decrease in Gini impurity or MSE in the resulting partitions). In this case, all we can do is choose a metric to *monitor* and use this metric of interest in selecting hyperparameters or comparing different methods (i.e., when *scoring* the method). The metric to be monitored is specified as the `scoring` parameter in the `cross_validate` function of `sklearn` or specified as an attribute in `model.compile` in `keras` models.

All available evaluation metrics in `sklearn` can be visualized (in alphabetical order) as the keys of the handy "SCORERS" dictionary:

```
print(sorted(sklearn.metrics.SCORERS.keys()))
```

Some common ones, together with some usage notes, are listed in Table 9.3.

Table 9.3: **Some common classification/regression metrics, together with some usage notes. The cross-entropy loss is named `log_loss` in `sklearn`.**

| Type | Name | Error/Score | Usage Notes |
|---|---|---|---|
| Classification | Accuracy | S | Not suitable for imbalanced data |
| | Precision | S | Yields high-purity result; false −s likely |
| | Recall | S | Yields complete result; false +s likely |
| | F1 score | S | Weighs the two type of errors similarly; possible alternative to accuracy |
| | Area under the curve | S | Useful for comparing algorithms |
| | Cross entropy | E | Suitable when predicting probabilities; rewards high-confidence correct predictions |
| Regression | $R^2$ score | S | Measures truth/predictions correlation; no direct interpretation |
| | MSE | E | Standard; sensitive to outliers |
| | RMSE | E | Like MSE, but easier to interpret |
| | MAE | E | Less skewed by outliers than MSE; returns "typical error" |

Evaluating unsupervised models is less trivial and, in general, more application dependent. For DR techniques, the most significant metric is often the loss of information in the reconstruction process. For example, let's say that we use some variant of PCA to reduce the dimensionality of some images. Once we project the images to their lower-dimensional PCA projection, they can't be compared with the original ones. However, we can *invert* the projection process (apply the inverse transformation to the objects in the lower dimensional space to go back to the higher dimensional space), and then compare the original images to the ones we have obtained. The difference between the images is a measure of the information lost as a result of the dimensionality reduction. Incidentally, this is exactly the idea behind autoencoders, as mentioned in Section 8.5.3. One important caveat is that all these measures of success tend to rely on the idea of difference between input and output, which is usually just the Euclidean distance between samples. This may or may not be a useful measure of information loss; for example, the Euclidean distance between two rotations of the same image may be very large, even if the amount of information loss is zero. As usual, understanding what matters for us and what's the best way of turning this idea into an evaluation metric is one of the most crucial parts of the process.

Other unsupervised learning applications, such as clustering, typically have their own evaluation procedures, from the elbow method to the silhouette score to the various information criteria for generative models discussed in Chapter 7.

If we want to compare two schemes for clustering, we can use the Adjusted Rand Index; to compare probability distributions, we could use the Earth Mover Distance or the Kullback-Leibler divergence. I will just remind you that these also deserve a careful assessment, as they are meaningful only if the method is a reasonable fit for the data. To be more practical, if you try to apply *k*-means to a distribution of data with nonconvex or nonglobular shaped clusters, the optimal number of clusters given by the elbow method will be meaningless. To complicate the matter, no obvious method exists to make sure that there exist convex or globular-shaped clusters. In other words, especially in unsupervised learning, it is very hard to answer the most important question of all: "Does it make sense to use this model here?"

### 9.1.4 Step Four: Benchmarking, diagnostics, optimization

You might now be feeling a bit frustrated, as we have just gone through a long process of data cleaning, problem setup, and evaluation, and yet we still have to run our first model! However, once we get to this point, I think the hard part is behind us.

For a supervised learning model, this section essentially corresponds to the pipeline established in Chapter 3 and shown in Figure 3.8. First, we decide whether our chosen method requires normalization/standardization of the data. Next, we choose a reference model (this could be, e.g., an instance of the chosen algorithm with default parameters), which will be the "benchmark," and we fit the data using a CV framework. This could also be a good time to assess whether we have sufficient computational resources to run CV models and hyperparameter optimization, as suggested in Table 9.1. If the answer is no, we have a few options. We could use one of the DR techniques described in Section 9.1.2 above, or we could work with a random subsample of the original data set for this part, after checking that the scores don't change significantly if we use the subset (otherwise we couldn't trust the optimization pipeline).

One important and tricky problem that can occur at this stage is: How do we decide what is a "good" score is? Should we be happy with 80% accuracy? Precision? Recall? Is an $R^2$ score of 0.8 good enough? Unfortunately, there is no universal answer to this question. All we can do is to follow good practices. We should try to establish expectations in the data exploration stage, for example, by plotting correlations between features and output, visualizing patterns, asking pertinent questions (e.g., "What would be the performance of a random classifier that only knows about class abundance?"), and being mindful about the potential pitfalls of our evaluation metric. There is no universal "good" score, but only the best score we can get given the data we have, and even in that case, there is no official test that can tell us whether we could improve further. "When to stop" will depend on what we need for our science, what the end goal is of our ML application, and how many resources we have. Even if the general idea of "best model" is the one with the highest test score, we might decide instead that interpretability, or robustness, or speed are more important to us. The advice that follows should be taken with a grain of salt and tailored to the specific problem being studied.

In the (very likely) event that the test scores are not 100% satisfactory, we can use a diagnostic tool, such as learning curves, for two purposes:

- To understand whether bias or variance are affecting our model, remembering that high bias is characterized by similar but poor train/test scores, and high variance is characterized by a statistically significant gap between train and test scores. The standard deviation of the test scores obtained through the CV process will help assess the statistical significance of the difference.
- To understand whether the current learning set size is posing a limitation to our model's performance, remembering that a significant residual slope observed in learning curves indicates that adding more training data would help, while learning curves that have plateaued indicate that the size of the learning set is not a significant driver of variance or bias.

The next step in this pipeline would be model optimization; typically by using hyperparameter tuning (see Section 3.6.1). As discussed in Chapters 3 and 5, the bias/variance decomposition can be a useful framework for understanding whether our model is excessively simplistic (high bias) or too capable of absorbing information, including noise (high variance), but we have also learned in Chapter 5 that it is not straightforward to interpret such information in terms of model complexity. Therefore, we should use our best judgment in choosing which hyperparameters may be worth tuning and which range of values is worth trying. We should do this rigorously by varying hyperparameters simultaneously (as they may be correlated), using one of the hyperparameter space exploration tools we have discussed, such as grid search or random search (see Section 4.5.1). However, I usually find it useful to "play around" with the hyperparameters first, one at a time, to build some intuition about which ones might be relevant and which ones tend to have a minimal effect. Finally, we should keep in mind that grid or random searches are used for the purpose of finding optimal parameters, but if we need to report fair generalization scores, we should use nested cross validation, as discussed in Section 4.5.3.

If we are dealing with an unsupervised learning problem, optimization might look different from the pipeline described above. However, the general framework is still valid: for example, we might optimize hyperparameters such as the number of clusters in a clustering scheme, or the number of components or neurons in a DR scheme. While we don't often hear about CV with respect to hyperparameter tuning in unsupervised learning methods, it actually makes sense to ensure that any hyperparameter optimization yields stable results on different partitions of the data.

### 9.1.5 Step Five: Assessment and further planning

I have titled this section "Step Five," but this stage is the most vague in our proposed pipeline. It consists of putting together all the information we have collected so far and using it to decide what to do next. Are we satisfied with our test scores and "done" with the project? If we aren't, which strategy would provide the best return on our effort, given what we have learned and the resources that we have? Does it make sense to switch gears and try a different algorithm? Would collecting new data be the most promising avenue to achieve our goal? Is computing power the most significant impediment to further improvement? These are a few examples of questions to ask at this point, and just like all worthwhile questions in science, they are truly open ended.

## 9.2 ADDITIONAL RESOURCES

This section is a (very limited) compilation of additional resources to continue your exploration of ML and deep learning methods in physics and astronomy.

I start with pedagogical resources, such as books, online classes, or classes whose materials are available. I have benefited from many of them.

In terms of foundational ML books written by and for scientists, my favorites are *Python Data Science Handbook* [VanderPlas, 2016], *Hands-on Machine Learning with Scikit-Learn, Keras, and TensorFlow* [Géron, 2019], and of course *Statistics, Data Mining, and Machine Learning in Astronomy* [Ivezić et al., 2014]. I found that for me, those texts struck just the right note in terms of being rigorous but also practical and easy to follow. A good deep learning textbook, although I am a bit less familiar with it, is *Deep Learning* [Goodfellow et al., 2016].

A fundamental reference in statistics and beyond is of course *The Elements of Statistical Learning* [Hastie et al., 2001], followed by the more recent and practical *An Introduction to Statistical Learning: with Applications in R* [James et al., 2013]. I also enjoyed reading parts of another classic, *Pattern Recognition and Machine Learning* [Bishop, 2006] which is a bit more formal but very well written. The draft book *Advanced Data Analysis from an Elementary Point of View*[Shalizi, in prep] was very helpful to me in many occasions as I was writing this book. Recently, I came across the book *Probability for Data Science* [Chan, 2021], which looks both useful and accessible, and is available for free (although I recommend supporting the author, if you have the means).

As far as online courses go, I started my own journey by taking the "classic" machine learning class[3] from Andrew Ng on Coursera, and I still think that it's hard to beat as an introductory class. I also have my own free online course that follows this book. It is available for free on the Flatiron's Open Learning server,[4] and I will

---

3 https://www.coursera.org/learn/machine-learning

4 https://apps.openlearning.flatironinstitute.org/learning/course/course-v1:cca+ML_01+A/home

make sure to add any changes on the book's website. I have used materials from, and heard great things about, Sebastian Raschka's machine learning classes[5] at the University of Wisconsin, which offer a good variety of levels, and Andreas Müller's Applied Machine Learning class at Columbia University, here.[6] Finally, the website "Machine Learning for Physicists"[7] contains class materials that are specialized for physics practitioners and are very good.

There are, of course, many learning platforms and a myriad of classes that are offered online; I often recommend asking fellow students to suggest some that can work for you, or using Twitter, if that's your thing, to find relevant threads, such as this one.[8]

### 9.2.1 Computational resources

As already mentioned in Chapter 8, one of the reasons machine learning and deep learning tools have become so widely used is improved *accessibility*: the existence of package managers such as `pip` or `anaconda`, libraries like `sklearn` [Pedregosa et al., 2011], deep learning frameworks such as `PyTorch` [Paszke et al., 2019], `TensorFlow` [Abadi et al., 2015], and high-level application programming interfaces such as `keras`, have greatly simplified the experience of building and optimizing machine learning models. Furthermore, training models on relatively small data sets is now possible on most personal laptops, or by making use of freely available resources such as Google Colab,[9] which also offers Graphics Processing Units and Tensor Processing Units, albeit the free version only allows short idle times.

Additional computational resources on the cloud are offered by several platforms, from Amazon Web Services (AWS) to Microsoft Azure to Google's Cloud Platform. They are not free, but there are programs for researchers to apply for grants for cloud computing credits, and it is also possible to use grant funding, when available, to pay for them.

### 9.2.2 Applications of ML in physics and astronomy

Deep learning models have become immensely popular in physics and astronomy, because of their ability to generate meaningful representations of high-dimensional spaces, their role in anomaly detection, and their usefulness as simulations toolboxes. It's impossible to come up with an even vaguely complete reference list, but I find that the NeurIPS "Machine Learning × Physical Sciences" Workshop, which was held for the first time in 2017 and then annually since 2019, provides a nice snapshot of how machine learning and deep learning techniques are used in the

---

5 https://sebastianraschka.com/teaching/

6 http://www.cs.columbia.edu/~amueller/

7 https://machine-learning-for-physicists.org/

8 https://twitter.com/marktenenholtz/status/1518560536555782144

9 https://colab.research.google.com/

field. The 2017 edition hosted 24 papers, which became 91 in 2019, around 150 in 2020 and 2021 and around 250 in 2022. All the papers are accessible from the main workshop site here.[10] Particle physicists also do a great job of keeping track of papers using ML in the "living document" [Feickert and Nachman, 2021], which also contains many useful references and examples of data sets and problems. Finally, the recently published book "Artificial Intelligence for High-Energy Physics" [Calafiura et al., 2022] contains a great collection of practical examples.

In the astronomy-specific landscape, I recommend keeping an eye on the website of the "Machine Learning for Astrophysics" workshop[11] at the International Conference of Machine Learning, which will hopefully become an annual appointment, and on the recently developed "Hello Universe" website,[12] an online live repository of Astronomy-related data sets and tutorials for machine learning and deep learning applications.

## 9.3 CONCLUSION

It is with great relief and the tiniest touch of sadness that I conclude this journey. As I mentioned in the Preface, I am painfully aware of the limitations of my own understanding of machine learning methods and tools, and of my ability to explain them. Nonetheless, I can proudly say that I gave it my best effort. I hope that this book contributed to make machine learning tools more accessible and less mysterious, and helped you progress in your own ML journey, whatever that might look like. In the words of my favorite poet, Eugenio Montale, translated by Jonathan Galassi and adapted by me:

*E senti allora,*
*se pure ti ripetono che puoi*
*fermarti a mezza via o in alto mare,*
*che non c'è sosta per noi,*
*ma strada, ancora strada,*
*e che il cammino è sempre da ricominciare.*

*And you feel then*
*even if they keep saying you can halt*
*halfway, or on the sea,*
*that there's no stopping for us,*
*but road and then more road,*
*and that the journey must always start anew.*

---

10 https://ml4physicalsciences.github.io/
11 https://ml4astro.github.io/icml2022/
12 https://archive.stsci.edu/hello-universe

# References


[Aad et al., 2012] Aad, G., Abajyan, T., Abbott, B., Abdallah, J. et al. (2012). Observation of a new particle in the search for the Standard Model Higgs boson with the ATLAS detector at the LHC. *Physics Letters B*, 716:1–29.

[Abadi et al., 2015] Abadi, M., Agarwal, A., Barham, P., Brevdo, E. et al. (2015). TensorFlow: Large-scale machine learning on heterogeneous systems. Software available from tensorflow.org.

[Acquaviva, 2019] Acquaviva, V. (2019). Pushing the technical frontier: From overwhelmingly large data sets to machine learning. *Proceedings of the International Astronomical Union*, 15(S341), 88–98. doi:10.1017/S1743921319003077.

[Acquaviva et al., 2020] Acquaviva, V., Lovell, C., and Ishida, E. (2020). Debunking generalization error or: How I learned to stop worrying and love my training set. *Proceedings of the International Astronomical Union*, 15(S341), 88–98. doi:10.1017/S174392131 9003077.

[Akaike, 1998] Akaike, H. (1998). Information theory and an extension of the maximum likelihood principle. In *Selected Papers of Hirotugu Akaike*, pages 199–213. New York: Springer.

[Ankerst et al., 1999] Ankerst, M., Breunig, M. M., Peter Kriegel, H., and Sander, J. (1999). Optics: Ordering Points to Identify the Clustering Structure, pages 49–60. New York: ACM Press.

[ATLAS Collaboration, 2021] ATLAS Collaboration (2021). Measurement of the $t\bar{t}t\bar{t}$ production cross section in *pp* collisions at $\sqrt{s}$=13 TeV with the ATLAS detector. *Journal of High Energy Physics* 11:1–53.

[Bańczyk et al., 2011] Bańczyk, K., Kempa, O., Lasota, T., and Trawiński, B. (2011). Empirical comparison of bagging ensembles created using weak learners for a regression problem. In *Asian Conference on Intelligent Information and Database Systems*, pages 312–322. Berlin: Springer.

[Baron, 2019] Baron, D. (2019). Machine learning in astronomy: A practical overview. *arXiv preprint arXiv:1904.07248*.

[Baydin et al., 2018] Baydin, A. G., Pearlmutter, B. A., Radul, A. A., and Siskind, J. M. (2018). Automatic differentiation in machine learning: A survey. *Journal of Machine Learning Research*, 18:11–43.

[Belkin et al., 2019] Belkin, M., Hsu, D., Ma, S., and Mandal, S. (2019). Reconciling modern machine-learning practice and the classical bias–variance trade-off. *Proceedings of the National Academy of Sciences*, 116(32):15849–15854.

[Ben-Hur and Weston, 2010] Ben-Hur, A., and Weston, J. (2010). A user's guide to support vector machines. In *Data Mining Techniques for the Life Sciences*, pages 223–239. New York: Springer.

[Bennett and Bredensteiner, 2010] Bennett, K. P., and Bredensteiner, E. J. (2010). Duality and geometry in SVM classifiers. In *Proceedings of the Seventeenth International Conference on Machine Learning*, pages 57–64.

[Bergstra et al., 2011] Bergstra, J., Bardenet, R., Bengio, Y., and Kégl, B. (2011). Algorithms for hyper-parameter optimization. In *Proceedings of Advances in Neural Information Processing Systems*, 24.

[Bergstra and Bengio, 2012] Bergstra, J., and Bengio, Y. (2012). Random search for hyper-parameter optimization. *Journal of Machine Learning Research*, 13(2): 281–305.

[Bishop, 2006] Bishop, Christopher M. (2006). *Pattern Recognition and Machine Learning*. New York: Springer.

[Boser et al., 1992] Boser, B. E., Guyon, I. M., and Vapnik, V. N. (1992). A training algorithm for optimal margin classifiers. In *Proceedings of the Fifth Annual Workshop on Computational Learning Theory*, COLT '92, pages 144–152. New York: Association for Computing Machinery.

[Bothwell et al., 2016] Bothwell, M. S., Maiolino, R., Cicone, C., Peng, Y., and Wagg, J. (2016). Galaxy metallicities depend primarily on stellar mass and molecular gas mass. *Astronomy & Astrophysics*, 595:A48.

[Brammer et al., 2012] Brammer, G. B., van Dokkum, P. G., Franx, M., Fumagalli, M., et al. (2012). 3D-HST: A wide-field grism spectroscopic survey with the Hubble Space Telescope. *The Astrophysical Journal Supplement Series*, 200:13.

[Brochu et al., 2010] Brochu, E., Cora, V. M., and De Freitas, N. (2010). A tutorial on Bayesian optimization of expensive cost functions, with application to active user modeling and hierarchical reinforcement learning. *arXiv preprint arXiv:1012.2599*.

[Brooijmans et al., 2020] Brooijmans, G., Buckley, A., Caron, S., Falkowski, A., et al. (2020). Les Houches 2019 physics at TeV colliders: New physics working group report. *arXiv:2002.12220*.

[Calafiura et al., 2022] Calafiura, P., Rousseau, D., and Terao, K. (2022). *Artificial Intelligence for High Energy Physics*. Singapore: World Scientific.

[Chan, 2021] Chan, S. H. (2021). *Introduction to Probability for Data Science*. Ann Arbor, MI: Michigan Publishing.

[Chatrchyan et al., 2012] Chatrchyan, S. et al. (2012). Observation of a new boson at a mass of 125 GeV with the CMS experiment at the LHC. *Physics Letters B*, 716:30–61.

[Cho et al., 2014] Cho, K., Van Merriënboer, B., Gulcehre, C., Bahdanau, D., et al. (2014). Learning phrase representations using rnn encoder-decoder for statistical machine translation. *arXiv preprint arXiv:1406.1078*.

[Chollet et al., 2015] Chollet, F., et al. (2015). Keras. https://keras.io.

[Chormunge and Jena, 2018] Chormunge, S., and Jena, S. (2018). Correlation based feature selection with clustering for high dimensional data. *Journal of Electrical Systems and Information Technology*, 5(3):542–549.

[Cooper et al., 2011] Cooper, M. C., Griffith, R. L., Newman, J. A., Coil, A. L., et al. (2011). The Deep3 galaxy redshift survey: The impact of environment on the size evolution of massive early-type galaxies at intermediate redshift. *Monthly Notices of the Royal Astronomical Society*, 419(4):3018–3027.

[Cover and Hart, 1967] Cover, T. M., and Hart, P. E. (1967). Nearest neighbor pattern classification. *IEEE Transactions on Information Theory*, 13:21–27.

[Cranmer et al., 2020] Cranmer, K., Brehmer, J., and Louppe, G. (2020). The frontier of simulation-based inference. *Proceedings of the National Academy of Sciences*, 117(48):30055–30062.

[Cybenko, 1989] Cybenko, G. (1989). Approximation by superpositions of a sigmoidal function. *Mathematics of Control, Signals and Systems*, 2(4):303–314.

[de Oliveira et al., 2016] de Oliveira, L., Kagan, M., Mackey, L., Nachman, B., & Schwartzman, A. (2016). Jet-images—deep learning edition. *Journal of High Energy Physics*, 2016(7), 1–32.

[Der Kiureghian and Ditlevsen, 2009] Der Kiureghian, A., and Ditlevsen, O. (2009). Aleatory or epistemic? Does it matter? *Structural Safety*, 31(2):105–112.

[Dieleman et al., 2015] Dieleman, S., Willett, K. W., and Dambre, J. (2015). Rotation-invariant convolutional neural networks for galaxy morphology prediction. *Monthly Notices of the Royal Astronomical Society*, 450:1441–1459.

[Dietterich, 2000] Dietterich, T. G. (2000). Ensemble methods in machine learning. In *International Workshop on Multiple Classifier Systems*, pages 1–15. New York: Springer.

[Doersch, 2016] Doersch, C. (2016). Tutorial on variational autoencoders. *arXiv preprint arXiv:1606.05908.*

[Drucker, 1997] Drucker, H. (1997). Improving regressors using boosting techniques. In *Proceedings of the International Conference on Machine Learning Research*, 97, pages 107–115.

[Duan et al., 2020] Duan, T., Anand, A., Ding, D. Y., Thai, K. K., et al. (2020). Ngboost: Natural gradient boosting for probabilistic prediction. In *Proceedings of the International Conference on Machine Learning*, pages 2690–2700.

[Duchi et al., 2011] Duchi, J., Hazan, E., and Singer, Y. (2011). Adaptive subgradient methods for online learning and stochastic optimization. *Journal of Machine Learning Research*, 12(7):2121–2159.

[Dumoulin and Visin, 2016] Dumoulin, V., and Visin, F. (2016). A guide to convolution arithmetic for deep learning. *arXiv preprint arXiv:1603.07285.*

[Ester et al., 1996] Ester, M., Kriegel, H.-P., Sander, J., and Xu, X. (1996). A density-based algorithm for discovering clusters in large spatial databases with noise. In *Proceedings of the Second International Conference on Knowledge Discovery and Data Mining*, pages 226–231.

[Farina et al., 2020] Farina, M., Nakai, Y., and Shih, D. (2020). Searching for new physics with deep autoencoders. *Physical Review D*, 101(7):075021.

[Fawcett, 2006] Fawcett, T. (2006). An introduction to ROC analysis. *Pattern Recognition Letters*, 27(8):861–874.

[Feickert and Nachman, 2021] Feickert, M., and Nachman, B. (2021). A living review of machine learning for particle physics. arXiv preprint arXiv:2102.02770.

[Florios et al., 2018] Florios, K., Kontogiannis, I., Park, S.-H., Guerra, J. A., et al. (2018). Forecasting solar flares using magnetogram-based predictors and machine learning. *Solar Physics*, 293(2):1–42.

[Fortmann-Roe, 2012] Fortmann-Roe, S. (2012). Understanding the bias-variance tradeoff. https://scott.fortmann-roe.com/docs/BiasVariance.html.

[Freund and Schapire, 1997] Freund, Y., and Schapire, R. E. (1997). A decision-theoretic generalization of on-line learning and an application to boosting. *Journal of Computer and System Sciences*, 55(1):119–139.

[Friedman, 2001] Friedman, J. H. (2001). Greedy function approximation: A gradient boosting machine. *Annals of Statistics*, 29(5):1189–1232.

[Gaillard et al., 1999] Gaillard, M. K., Grannis, P. D., and Sciulli, F. J. (1999). The standard model of particle physics. *Reviews of Modern Physics*, 71(2):S96.

[Gal and Ghahramani, 2016] Gal, Y., and Ghahramani, Z. (2016). Dropout as a Bayesian approximation: Representing model uncertainty in deep learning. In *Proceedings of the International Conference on Machine Learning*, pages 1050–1059.

[Gawlikowski et al., 2021] Gawlikowski, J., Rovile Njieutcheu Tassi, C., Ali, M., Lee, J., et al. (2021). A survey of uncertainty in deep neural networks. *arXiv preprint arXiv:2107.03342*.

[Geman et al., 1992] Geman, S., Bienenstock, E., and Doursat, R. (1992). Neural networks and the bias/variance dilemma. *Neural Computation*, 4(1):1–58.

[Géron, 2019] Géron, A. (2019). *Hands-on Machine Learning with Scikit-Learn, Keras, and TensorFlow: Concepts, Tools, and Techniques to Build Intelligent Systems*. Sebastopol, CA: O'Reilly Media.

[Geurts, 2009] Geurts, P. (2009). Bias vs variance decomposition for regression and classification. In *Data Mining and Knowledge Discovery Handbook*, pages 733–746. Boston: Springer.

[Glorot and Bengio, 2010] Glorot, X., and Bengio, Y. (2010). Understanding the difficulty of training deep feedforward neural networks. In *Proceedings of the Thirteenth International Conference on Artificial Intelligence and Statistics (AISTATS'10)*, pages 249–256. Society for Artificial Intelligence and Statistics.

[Goodfellow et al., 2016] Goodfellow, I., Bengio, Y., and Courville, A. (2016). *Deep Learning*. Cambridge, MA: MIT Press. http://www.deeplearningbook.org.

[Goodfellow et al., 2014] Goodfellow, I. J., Pouget-Abadie, J., Mirza, M., Xu, B., et al. (2014). Generative adversarial networks. *Communications of the ACM*, 63(11):139–144.

[Greif and Lannon, 2020] Greif, K., and Lannon, K. (2020). Physics inspired deep neural networks for top quark reconstruction. *EPJ Web of Conferences*, 245:06029.

[Hastie et al., 2001] Hastie, T., Tibshirani, R., and Friedman, J. (2001). *The Elements of Statistical Learning*. Springer Series in Statistics. New York: Springer.

[He et al., 2015] He, K., Zhang, X., Ren, S., and Sun, J. (2015). Delving deep into rectifiers: Surpassing human-level performance on imagenet classification. In *Proceedings of the IEEE International Conference on Computer Vision*, pages 1026–1034.

[Hinton et al., 2020] Hinton, G., Srivastava, N., and Swersky, K. (2020). Neural networks for machine learning lecture 6a overview of mini-batch gradient descent. http://www.cs.toronto.edu/~hinton/coursera/lecture6/lec6.pdf.

[Hochreiter and Schmidhuber, 1997] Hochreiter, S., and Schmidhuber, J. (1997). Long short-term memory. *Neural Computation*, 9(8):1735–1780.

[Hocking et al., 2018] Hocking, A., Geach, J. E., Sun, Y., and Davey, N. (2018). An automatic taxonomy of galaxy morphology using unsupervised machine learning. *Monthly Notices of the Royal Astronomical Society*, 473:1108–1129.

[Hogg and Villar, 2021] Hogg, D. W., and Villar, S. (2021). Fitting very flexible models: Linear regression with large numbers of parameters. *Publications of the Astronomical Society of the Pacific*, 133(1027): 093001.

[Hogg et al., 2010] Hogg, D. W., Bovy, J., and Lang, D. (2010). Data analysis recipes: Fitting a model to data. *arXiv e-prints*, arXiv:1008.4686.

[Hornik et al., 1989] Hornik, K., Stinchcombe, M., White, H., et al. (1989). Multilayer feedforward networks are universal approximators. *Neural Networks*, 2(5):359–366.

[Hüllermeier and Waegeman, 2021] Hüllermeier, E., and Waegeman, W. (2021). Aleatoric and epistemic uncertainty in machine learning: An introduction to concepts and methods. *Machine Learning*, 110(3):457–506.

[Ioffe and Szegedy, 2015] Ioffe, S., and Szegedy, C. (2015). Batch normalization: Accelerating deep network training by reducing internal covariate shift. In *Proceedings of the International Conference on Machine Learning*, pages 448–456.

[Ising, 1925] Ising, E. (1925). Beitrag zur Theorie des Ferromagnetismus. *Zeitschrift für Physik*, 31(1):253–258.

[Ivezić et al., 2014] Ivezić, Ž., Connolly, A. J., VanderPlas, J. T., and Gray, A. (2014). *Statistics, Data Mining, and Machine Learning in Astronomy: A Practical Python Guide for the Analysis of Survey Data*, vol. 1. Princeton, NJ: Princeton University Press.

[Ivezić et al., 2019] Ivezić, Ž., Kahn, S. M., Tyson, J. A., Abel, B., et al. (2019). LSST: From science drivers to reference design and anticipated data products. *Astrophysical Journal*, 873(2): 111.

[James et al., 2013] James, G., Witten, D., Hastie, T., and Tibshirani, R. (2013). *An Introduction to Statistical Learning*, vol. 112. New York: Springer.

[James, 2003] James, G. M. (2003). Variance and bias for general loss functions. In *Machine Learning*, 51(2): 115–135.

[Ji et al., 2019] Ji, X., Henriques, J. F., and Vedaldi, A. (2019). Invariant information clustering for unsupervised image classification and segmentation. In *Proceedings of the IEEE/CVF International Conference on Computer Vision*, pages 9865–9874. IEEE.

[Kearns and Valiant, 1994] Kearns, M., and Valiant, L. (1994). Cryptographic limitations on learning boolean formulae and finite automata. *Journal of the ACM*, 41(1):67–95.

[Kepler] Kepler and k2 website. https://www.nasa.gov/mission_pages/kepler/overview /index.html.

[Kingma and Ba, 2014] Kingma, D. P., and Ba, J. (2014). Adam: A method for stochastic optimization. *arXiv preprint arXiv:1412.6980*.

[Kingma and Welling, 2013] Kingma, D. P., and Welling, M. (2013). Auto-encoding variational Bayes. *arXiv preprint arXiv:1312.6114*.

[Kohonen, 1997] Kohonen, T. (1997). Exploration of very large databases by self-organizing maps. In *Proceedings of the International Conference on Neural Networks (ICNN'97)*, vol. 1, pages PL1–PL6. IEEE.

[Komiske et al., 2019] Komiske, P. T., Metodiev, E. M., and Thaler, J. (2019). Metric space of collider events. *Physical Review Letters*, 123(4).

[Kopparapu et al., 2014] Kopparapu, R. K., Ramirez, R. M., SchottelKotte, J., Kasting, J. F., et al. (2014). Habitable zones around main-sequence stars: Dependence on planetary mass. *Astrophysical Journal Letters*, 787(2):L29.

[Krizhevsky and Hinton, 2009] Krizhevsky, A., and Hinton, G. (2009). Learning multiple layers of features from tiny images. Master's thesis, University of Tront.

[Lakshminarayanan et al., 2016] Lakshminarayanan, B., Pritzel, A., and Blundell, C. (2016). Simple and scalable predictive uncertainty estimation using deep ensembles. Proceedings of the 31st International Conference on Neural Information Processing Systems, NIPS'17, pages 6405–6416.

[Le Cun et al., 1989] Le Cun, Y., Boser, B., Denker, J. S., Henderson, D., et al. (1989). Handwritten digit recognition with a back-propagation network. In *Proceedings of the 2nd International Conference on Neural Information Processing Systems*, NIPS'89, pages 396–404. Cambridge, MA: MIT Press.

[LeCun et al., 1998] LeCun, Y., Bottou, L., Orr, G. B., and Müller, K.-R. (1998). Efficient backprop. In *Neural Networks: Tricks of the Trade*, pages 9–49. Berlin: Springer.

[Little and Rubin, 1986] Little, R. J. A., and Rubin, D. B. (1986). *Statistical Analysis with Missing Data*. Hoboken, NJ: John Wiley & Sons.

[Louppe, 2014] Louppe, G. (2014). Understanding random forests: From theory to practice. *arXiv preprint arXiv:1407.7502*.

[Lundberg and Lee, 2017] Lundberg, S. M., and Lee, S.-I. (2017). A unified approach to interpreting model predictions. In Guyon, I., Luxburg, U. V., Bengio, S., Wallach, H., et al., editors,

*Advances in Neural Information Processing Systems 30*, pages 4765–4774. Red Hook, NY: Curran Associates.

[McCulloch and Pitts, 1943] McCulloch, W. S., and Pitts, W. (1943). A logical calculus of the ideas immanent in nervous activity. *Bulletin of Mathematical Biophysics*, 5(4):115–133.

[McKinney, 2010] McKinney, W. (2010). Data structures for statistical computing in Python. In van der Walt, S., and Millman, J., editors, *Proceedings of the 9th Python in Science Conference*, pages 56–61.

[Megahed et al., 2020] Megahed, F. M., Jones-Farmer, L. A., and Rigdon, S. E. (2020). A retrospective cluster analysis of covid-19 cases by county. https://doi.org/10.1101/2020.11.12.379537.

[Mehta et al., 2019] Mehta, P., Bukov, M., Wang, C.-H., Day, A. G., et al. (2019). A high-bias, low-variance introduction to machine learning for physicists. *Physics Reports*, 810: 1–124.

[Meir and Rätsch, 2003] Meir, R., and Rätsch, G. (2003). An introduction to boosting and leveraging. In *Advanced Lectures on Machine Learning*, pages 118–183. New York: Springer.

[Morningstar and Melko, 2018] Morningstar, A., and Melko, R. G. (2018). Deep learning the Ising model near criticality. *Journal of Machine Learning Research*, 18:1–17.

[Morvan et al., 2021] Morvan, M. L., Josse, J., Scornet, E., and Varoquaux, G. (2021). What's a good imputation to predict with missing values? *Advances in Neural Information Processing Systems*, 34:11530–11540.

[Mullachery et al., 2018] Mullachery, V., Khera, A., and Husain, A. (2018). Bayesian neural networks. *arXiv:1801.07710*.

[Murphy and Maggioni, 2018] Murphy, J. M., and Maggioni, M. (2018). Unsupervised clustering and active learning of hyperspectral images with nonlinear diffusion. *IEEE Transactions on Geoscience and Remote Sensing*, 57(3):1829–1845.

[Nachman, 2020] Nachman, B. (2020). Anomaly detection for physics analysis and less than supervised learning. *arXiv preprint arXiv:2010.14554*.

[Natekin and Knoll, 2013] Natekin, A., and Knoll, A. (2013). Gradient boosting machines, a tutorial. *Frontiers in Neurorobotics*, 7:21.

[Neal et al., 2018] Neal, B., Mittal, S., Baratin, A., Tantia, V., et al. (2018). A modern take on the bias-variance tradeoff in neural networks. *arXiv preprint arXiv:1810.08591*.

[Newman et al., 2013] Newman, J. A., Cooper, M. C., Davis, M., Faber, S. M., et al. (2013). The DEEP2 galaxy redshift survey: Design, observations, data reduction, and redshifts. *Astrophysical Journal Supplement Series*, 208(1):5.

[Ng, 2019] Ng, A. (2019). CS 229 notes, https://see.stanford.edu/materials/aimlcs229/cs229-notes3.pdf

[Nielsen, 2018] Nielsen, M. A. (2018). *Neural Networks and Deep Learning* (vol. 25). San Francisco: Determination Press.

[Opitz and Maclin, 1999] Opitz, D., and Maclin, R. (1999). Popular ensemble methods: An empirical study. *Journal of Artificial Intelligence Research*, 11:169–198.

[Pascanu et al., 2013] Pascanu, R., Mikolov, T., and Bengio, Y. (2013). On the difficulty of training recurrent neural networks. In *Proceedings of the International Conference on Machine Learning*, vol. 28, pages 1310–1318.

[Paszke et al., 2019] Paszke, A., Gross, S., Massa, F., Lerer, A., et al. (2019). Pytorch: An imperative style, high-performance deep learning library. In Wallach, H., Larochelle, H., Beygelzimer, A., d'Alché-Buc, F., et al., editors, *Advances in NeurIPS, 32*, pages 8024–8035. Red Hook, NY: Curran Associates.

[Pedregosa et al., 2011] Pedregosa, F., Varoquaux, G., Gramfort, A., Michel, V., et al. (2011). Scikit-learn: Machine learning in Python. *Journal of Machine Learning Research*, 12:2825–2830.

[Portillo et al., 2020] Portillo, S. K., Parejko, J. K., Vergara, J. R., and Connolly, A. J. (2020). Dimensionality reduction of SDSS spectra with variational autoencoders. *Astronomical Journal*, 160(1):45.

[Probst et al., 2020] Probst, P., Boulesteix, A.-L., and Bischl, B. (2019). Tunability: Importance of hyperparameters of machine learning algorithms. *Journal of Machine Learning Research*, 20(53):1–32.

[Rubner et al., 2000] Rubner, Y., Tomasi, C., and Guibas, L. J. (2000). The earth mover's distance as a metric for image retrieval. *International Journal of Computer Vision*, 40(2):99–121.

[Rumelhart et al., 1985] Rumelhart, D. E., Hinton, G. E., and Williams, R. J. (1985). Learning internal representations by error propagation. Technical report, University of California, San Diego, La Jolla Institute for Cognitive Science.

[Salvato et al., 2019] Salvato, M., Ilbert, O., and Hoyle, B. (2019). The many flavours of photometric redshifts. *Nature Astronomy*, 3(3):212–222.

[Schapire, 1990] Schapire, R. E. (1990). The strength of weak learnability. *Machine Learning*, 5(2):197–227.

[Schölkopf and Smola, 2001] Schölkopf, B., and Smola, A. J. (2001). *Learning with Kernels: Support Vector Machines, Regularization, Optimization, and Beyond*. Cambridge, MA: MIT Press.

[Schölkopf et al., 1999] Schölkopf, B., Smola, A. J., and Müller, K.-R. (1999). *Kernel Principal Component Analysis*, pages 327–352. Cambridge, MA: MIT Press.

[Schwartz, 2018] Schwartz, M. D. (2018). *TASI Lectures on Collider Physics*, pages 65–100. In *Anticipating the Next Discoveries in Particle Physics: TASI 2016 Proceedings of 2016 Theoretical Advanced Study Institute in Elementary Particle Physics* (pp. 65–100).

[Schwarz, 1978] Schwarz, G. (1978). Estimating the dimension of a model. *The Annals of Statistics*, 6(2), 461–464.

[Shalizi, in prep] Shalizi, C. (in prep.). *Advanced Data Analysis from an Elementary Point of View*. Oxford: Oxford University Press.

[Sherstinsky, 2020] Sherstinsky, A. (2020). Fundamentals of recurrent neural network (RNN) and long short-term memory (LSTM) network. *Physica D: Nonlinear Phenomena*, 404:132306.

[Snoek et al., 2012] Snoek, J., Larochelle, H., and Adams, R. P. (2012). Practical Bayesian optimization of machine learning algorithms. In *Advances in Neural Information Processing Systems 25* (NIPS 2012) Edited by: F. Pereira and C. J. Burges and L. Bottou and K. Q. Weinberger

[Sorzano et al., 2014] Sorzano, C. O. S., Vargas, J., and Montano, A. P. (2014). A survey of dimensionality reduction techniques. *arXiv preprint*, arXiv:1403.2877.

[Stekhoven and Bühlmann, 2012] Stekhoven, D. J., and Bühlmann, P. (2012). Missforest—non-parametric missing value imputation for mixed-type data. *Bioinformatics*, 28(1): 112–118.

[Sutera et al., 2021] Sutera, A., Louppe, G., Huynh-Thu, V. A., Wehenkel, L., and Geurts, P. (2021). From global to local MDI variable importances for random forests and when they are Shapley values. *Advances in Neural Information Processing Systems*, 34, 3533–3543. *arXiv preprint arXiv:2111.02218*.

[TESS] Tess website. https://www.nasa.gov/tess-transiting-exoplanet-survey-satellite.

[Troyanskaya et al., 2001] Troyanskaya, O., Cantor, M., Sherlock, G., Brown, P., et al. (2001). Missing value estimation methods for DNA microarrays. *Bioinformatics*, 17(6):520–525.

[van Buuren and Groothuis-Oudshoorn, 2011] van Buuren, S., and Groothuis-Oudshoorn, K. (2011). Mice: Multivariate Imputation by Chained Equations in R. *Journal of Statistical Software*, 45(3), 1–67.

[Vandenberg-Rodes et al., 2016] Vandenberg-Rodes, A., Moftakhari, H. R., AghaKouchak, A., Shahbaba, B., et al. (2016). Projecting nuisance flooding in a warming climate using generalized linear models and Gaussian processes. *Journal of Geophysical Research: Oceans*, 121(11):8008–8020.

[van der Maaten and Hinton, 2008] van der Maaten, L., and Hinton, G. (2008). Visualizing high-dimensional data using t-SNE. *Journal of Machine Learning Research*, 9(11).

[VanderPlas, 2016] VanderPlas, J. (2016). *Python Data Science Handbook: Essential Tools for Working with Data*. Sebastopol, CA: O'Reilly Media.

[Van Leeuwen, 1991] Van Leeuwen, J. (1991). *Handbook of Theoretical Computer Science (vol. A) Algorithms and Complexity*. Cambridge, MA: MIT Press.

[Veltman, 2018] Veltman, M. J. (2018). *Facts and Mysteries in Elementary Particle Physics*. Singapore: World Scientific.

[Wang and Yeung, 2020] Wang, H., and Yeung, D.-Y. (2020). A survey on Bayesian deep learning. *ACM Computing Surveys (CSUR)*, 53(5):1–37.

[Wu et al., 2019] Wu, Z., Shen, C., and Van Den Hengel, A. (2019). Wider or deeper: Revisiting the resnet model for visual recognition. *Pattern Recognition*, 90:119–133.

[Yadav et al., 2019] Yadav, H., Candela, A., and Wettergreen, D. (2019). A study of unsupervised classification techniques for hyperspectral datasets. In *International Geoscience and Remote Sensing Symposium*, pages 2993–2996. IEEE.

[Yu and Zhu, 2020] Yu, T., and Zhu, H. (2020). Hyper-parameter optimization: A review of algorithms and applications. *arXiv preprint arXiv:2003.05689*.

[Yuan, 2018] Yuan, L. (2018). Implementation of self-organizing maps with Python. Masters thesis. Kingston, RI: University of Rhode Island.

[Zadeh, 2015] Zadeh, R. (2015). Cme 323: Distributed algorithms and optimization. https://stanford.edu/~rezab/classes/cme323/S15/notes/lec11.pdf.

[Zhou et al., 2019] Zhou, R., Cooper, M. C., Newman, J. A., Ashby, M. L., et al. (2019). Deep ugrizy imaging and deep2/3 spectroscopy: A photometric redshift testbed for LSST and public release of data from the DEEP3 galaxy redshift survey. *Monthly Notices of the Royal Astronomical Society*, 488(4):4565–4584.

# Index